T0213653

Lecture Notes in Computer Science 10365

Commenced Publication in 1973
Founding and Former Series Editors:
Gerhard Goos, Juris Hartmanis, and Jan van Leeuwen

More information about this series at http://www.springer.com/series/7409

Andrea Calì · Peter Wood
Nigel Martin · Alexandra Poulovassilis (Eds.)

Data Analytics

31st British International Conference
on Databases, BICOD 2017
London, UK, July 10–12, 2017
Proceedings

 Springer

Editors
Andrea Calì
Birkbeck, University of London
London
UK

Nigel Martin
Birkbeck, University of London
London
UK

Peter Wood
Birkbeck, University of London
London
UK

Alexandra Poulovassilis
Birkbeck, University of London
London
UK

ISSN 0302-9743   ISSN 1611-3349 (electronic)
Lecture Notes in Computer Science
ISBN 978-3-319-60794-8   ISBN 978-3-319-60795-5   (eBook)
DOI 10.1007/978-3-319-60795-5

Library of Congress Control Number: 2017943849

LNCS Sublibrary: SL3 – Information Systems and Applications, incl. Internet/Web, and HCI

Printed on acid-free paper

This Springer imprint is published by Springer Nature
The registered company is Springer International Publishing AG
The registered company address is: Gewerbestrasse 11, 6330 Cham, Switzerland

Preface

This volume contains research papers presented at BICOD 2017: the 31st British International Conference on Databases held July 10–12, 2017, at Birkbeck, University of London.

The BICOD Conference is an international venue with a long tradition of presentation and discussion of research in the broad area of data management. The theme of BICOD 2017 was "Data Analytics," that is, the process of deriving higher-level information from large sets of raw data.

The conference featured keynotes from three distinguished speakers: Sihem Amer-Yahia, Laboratoire d'Informatique de Grenoble; Tim Furche, University of Oxford and Wrapidity; and Elena Baralis, Politecnico di Torino. There were also two invited tutorials: by Vasiliki Kalavri, ETH Zürich, on distributed graph processing; and by Leopoldo Bertossi, Carleton University, on declarative approaches to data quality assessment and cleaning.

BICOD 2017 attracted paper submissions from Australia, Austria, Canada, Colombia, Germany, Ireland, Italy, Russia, Tunisia, and the UK. The Program Committee accepted 16 papers to appear in the proceedings covering topics such as data cleansing, data integration, data wrangling, data mining and knowledge discovery, graph data and knowledge graphs, intelligent data analysis, approximate and flexible querying, data provenance, and ontology-based data access.

A prize was awarded for the best paper presented by a PhD student, which was generously sponsored by Neo Technology. The authors of a small number of selected best papers were invited to submit extended versions of their work to a special issue of *The Computer Journal*.

We would like to thank all the authors for submitting their work to BICOD, the Program Committee members for their contribution to selecting an inspiring set of research papers, and the distinguished invited speakers for accepting our invitation to present their work in London. We also gratefully thank everyone involved in the local organization, including Tara Orlanes-Angelopoulou, Phil Gregg, Lucy Tallentire, Catherine Griffiths, and Matthew Jayes, for all their hard work and dedication in making this conference possible.

May 2017

Andrea Calì
Peter Wood
Nigel Martin
Alexandra Poulovassilis

Organization

Program Committee

Andrea Calì	Birkbeck, University of London, UK (PC Co-chair)
Alvaro Fernandes	University of Manchester, UK
George Fletcher	Eindhoven University of Technology, The Netherlands
Floris Geerts	University of Antwerp, Belgium
Sven Helmer	Free University of Bozen-Bolzano, Italy
Anne James	Coventry University, UK
Mike Jackson	Birmingham City University, UK
Sebastian Maneth	University of Edinburgh, UK
Nigel Martin	Birkbeck, University of London, UK
Werner Nutt	Free University of Bozen-Bolzano, Italy
Dan Olteanu	University of of Oxford, UK
Andreas Pieris	University of Edinburgh, UK
Alessandro Provetti	Birkbeck, University of London, UK
Alexandra Poulovassilis	Birkbeck, University of London, UK
Mark Roantree	Dublin City University, Ireland
Maria-Esther Vidal	Universidad Simón Bolívar, Venezuela
John Wilson	University of Strathclyde, UK
Peter Wood	Birkbeck, University of London, UK (PC Co-chair)

Keynotes

Wrapping Millions of Documents Per Day - and How That's Just the Beginning

Tim Furche

Department of Computer Science, University of Oxford, Oxford, UK
CTO Wrapidity Ltd., Oxford, UK
tim@furche.net

Abstract. Companies and researchers have been painfully maintaining manually created programs for wrapping (aka scraping) web sites for decades - employing hundreds of engineers to wrap thousands or even tens of thousands of sources. Where wrappers break, engineers must scramble to fix the wrappers manually or face the ire of their users. In DIADEM we demonstrated the effectiveness of hybrid AIs for automatically generating wrapper programs in an academic settings. Our hybrid approach combined knowledge-based rule systems with large-scale analytics founded in techniques from the NLP and ML community.

Recently, we commercialized that technology into Wrapidity Ltd. and quickly joined up with Meltwater, the leading global media-intelligence company. At Meltwater, we are applying Wrapidity's techniques and insight to the wrapping of tens of thousands of news sources, processing over 10M unique documents per day. But that's just the beginning .... Together with Meltwater engineers and researchers from San Francisco, Stockholm, Bangalore, and Budapest, we are on the way to make the collection and analysis of "outside" data a breeze through fairhair.ai, a platform for accessing and enriching the vast content stored and collected by Meltwater every day.

Toward Interactive User Data Analytics

Sihem Amer-Yahia

CNRS, University Grenoble-Alps, Grenoble, France

Abstract. User data can be acquired from various domains. This data is characterized by a combination of demographics such as age and occupation and user actions such as rating a movie, reviewing a restaurant or buying groceries. User data is appealing to analysts in their role as data scientists who seek to conduct large-scale population studies, and gain insights on various population segments. It is also appealing to novice users in their role as information consumers who use the social Web for routine tasks such as finding a book club or choosing a restaurant.

User data exploration has been formulated as identifying group-level behavior such as *Asian women who publish regularly in databases*. Group-level exploration enables new findings and addresses issues raised by the peculiarities of user data such as noise and sparsity. I will review our work on one-shot and interactive user data exploration. I will then describe the challenges of developing a visual analytics tool for finding and connecting users and groups.

Opening the Black Box: Deriving Rules from Data

Elena Baralis

Dipartimento di Automatica e Informatica,
Politecnico di Torino, Torino, Italy
elena.baralis@polito.it

Abstract. A huge amount of data is currently being made available for exploration and analysis in many application domains. Patterns and models are extracted from data to describe their characteristics and predict variable values. Unfortunately, many high quality models are characterized by being hardly interpretable. Rules mined from data may provide easily interpretable knowledge, both for exploration and classification (or prediction) purposes.

In this talk I will introduce different types of rules (e.g., several variations on association rules, classification rules) and will discuss their capability of describing phenomena and highlighting interesting correlations in data.

Tutorials

Programming Models and Tools for Distributed Graph Processing

Vasiliki Kalavri

Systems Group, ETH Zürich, Zürich, Switzerland
kalavriv@inf.ethz.ch

1. Introduction

Graphs capture relationships between data items, such as interactions or dependencies, and their analysis can reveal valuable insights for machine learning tasks, anomaly detection, clustering, recommendations, social influence analysis, bioinformatics, and other application domains. Graph analysis is often performed in a distributed manner, where graph datasets are partitioned and processed on several machines. Distributed graph processing is practical for two reasons. First, it enables the analysis of massive graph datasets, such as social networks, particle simulations, web access history, product ratings, and others that might not fit in the memory of a single machine. Second, it provides a convenient way to build end-to-end data analysis pipelines. Graphs rarely appear as raw data; they are most often derived by transforming other data sets into graphs. Data entities of interest are extracted and modeled as graph nodes and their relationships are modeled as edges. Thus, graph representations frequently appear in an intermediate step of some larger distributed data processing pipeline. Such intermediate graph-structured data are already partitioned upon creation and, thus, distributed algorithms are essential in order to efficiently analyze them and avoid expensive data transfers.

Writing a distributed graph processing application is a challenging task that requires handling computation parallelization, data partitioning, and communication efficiently. Furthermore, graph applications are highly diverse and expose a variety of data access and communication patterns. For example, iterative refinement algorithms, like PageRank, can be expressed as parallel computations over the local neighborhood of each vertex. Graph traversal algorithms produce unpredictable access patterns, while graph aggregations require grouping of similar vertices or edges together. To address the challenges of distributed graph processing, several high-level programming abstractions and respective system implementations have been recently proposed [1–5]. Even though some have gained more popularity than others, each abstraction is optimized for certain classes of graph applications. For instance, the popular vertex-centric model is well-suitable for iterative value propagation algorithms, while the neighborhood-centric model is designed to efficiently support operations on custom subgraphs, like ego networks.

2. Goal of this Tutorial

This tutorial reviews the state of the art in high-level abstractions for distributed graph processing. First, we present six models that were developed specifically for distributed graph processing, namely vertex-centric, scatter-gather, gather-sum-apply-scatter, subgraph-centric, filter-process, and graph traversals. Then, we consider general-purpose distributed programming models that have been used for graph analysis, such as MapReduce, dataflow, linear algebra primitives, datalog, and shared partitioned tables.

The tutorial aims at making a *qualitative* comparison of popular graph programming abstractions. We examine each programming model with regard to semantics, expressiveness, and applicability. We identify classes of graph algorithms that can be naturally expressed by each abstraction and we give examples of application domains for which a model may appear to be non-intuitive. We use the PageRank algorithm as the running example throughout the tutorial to provide a common ground for comparison and demonstrate how different models can express a fairly simple and familiar graph computation. We further consider performance limitations of some graph programming models and we summarize proposed extensions and optimizations.

References

1. Gonzalez, J.E., Low, Y., Gu, H., Bickson, D., Guestrin, C.: PowerGraph: distributed graph-parallel computation on natural graphs. In: Proceedings of the 10th USENIX Symposium on Operating Systems Design and Implementation, pp. 17–30 (2012)
2. Low, Y., Bickson, D., Gonzalez, J., Guestrin, C., Kyrola, A., Hellerstein, J.M.: Distributed GraphLab: a framework for machine learning and data mining in the cloud. Proc. VLDB Endow. 5(8), 716–727 (2012)
3. Malewicz, G., Austern, M.H., Bik, A.J., Dehnert, J.C., Horn, I., Leiser, N., Czajkowski, G.: Pregel: a system for large-scale graph processing. In: Proceedings of the 2010 ACM SIGMOD International Conference on Management of Data, pp. 135–146. ACM (2010)
4. Stutz P., Bernstein A., Cohen W.: Signal/Collect: graph algorithms for the (semantic) web. In: Patel-Schneider P.F., et al. (eds.) ISWC 2010. LNCS, vol. 6496, pp. 764–780. Springer, Heidelberg (2010)
5. Tian, Y., Balmin, A., Corsten, S.A., Tatikonda, S., McPherson, J.: From think like a vertex to think like a graph. Proc. VLDB Endow. 7(3), 193–204 (2013)

Declarative Approaches to Data Quality Assessment and Cleaning

Leopoldo Bertossi

Carleton University, Ottawa, Canada
bertossi@scs.carleton.ca

Abstract. Data quality is an increasingly important problem and concern in business intelligence. Data quality in general, so too their cleaning and assessment in terms of quality in particular, are relative properties and activities, which largely depend on additional semantic information, data and metadata. This extra information and their use for data quality purposes can be specified in declarative terms. In this tutorial we review and discuss some forms of declarative semantic conditions that have their origin in integrity constraints, quality constraints, matching dependencies, and ontological contexts, etc. They can all be used to characterize, assess and obtain quality data from possibly dirty data sources. In this tutorial the emphasis is on declarative approaches to inconsistency, certain forms of incompleteness, and duplicate data. The latter give rise to the entity resolution problem, for which we show declarative approaches that can be naturally combined with machine learning methods.

Contents

Keynote

Toward Interactive User Data Analytics

Sihem Amer-Yahia[✉]

CNRS, University Grenoble-Alps, Grenoble, France
Sihem.Amer-Yahia@imag.fr

Abstract. User data can be acquired from various domains. This data is characterized by a combination of demographics such as age and occupation and user actions such as rating a movie, reviewing a restaurant or buying groceries. User data is appealing to analysts in their role as data scientists who seek to conduct large-scale population studies, and gain insights on various population segments. It is also appealing to novice users in their role as information consumers who use the social Web for routine tasks such as finding a book club or choosing a restaurant.

User data exploration has been formulated as identifying group-level behavior such as *Asian women who publish regularly in databases*. Group-level exploration enables new findings and addresses issues raised by the peculiarities of user data such as noise and sparsity. I will review our work on one-shot [1–4] and interactive [5] user data exploration. I will then describe the challenges of developing a visual analytics tool for finding and connecting users and groups.

1 Data Model

Given a set of users U and a set of items I, we define *user data* as a database D of tuples $\langle u, i, val \rangle$ which represent a value val induced by an action such as browsing, tagging, rating, and tweeting, of user u, from a set U, on item i, from a set I. For instance, the tuple $\langle John, Titanic, 4 \rangle$ describes that John rated the movie Titanic with a score of 4. The tuple $\langle Tiffany, tweet_{id}, Hemophilia \rangle$ describes that Tiffany tweets about 'Hemophilia'.

Users and items have attributes drawn from a set A. The set of user attributes (age, gender, diet, occupation, etc.) is denoted as $A_u \subset A$ and the set of item attributes (book author, movie director, tweet language, etc.) is denoted as A_i where $A_u \cup A_i = A$ and $A_i \cap A_u = \emptyset$. Each attribute $a_i \in A$ has a set of values $dom(a_i) = \{v_1^i, v_2^i \dots \}$. V denotes the set of all attribute values.

Multiple datasets could be represented in our model. We have been using over 5B tweets, about 300M customer receipts from a retail chain of 1,800 stores, 10 M rating records from MovieLens, 50 M artist ratings from LastFM, and about 200 K book ratings from BookCrossing.

User Group. A user group g, is a subset of U to which is associated a description defined as $[v_1^1, v_2^1 \dots v_j^i \dots]$ where each $v_j^i \in dom(a_i)$ either holds for all users in g (if $a_i \in A_u$) or holds for all items in g (if $a_i \in A_i$). For instance, the group $[25, student, action]$ contains 25-year old students who watch action

© Springer International Publishing AG 2017
A. Calì et al. (Eds.): BICOD 2017, LNCS 10365, pp. 3–6, 2017.
DOI: 10.1007/978-3-319-60795-5_1

movies. Here, "25" $\in dom(age)$, "student" $\in dom(occupation)$ and "action" $\in dom(genre)$ where $\{age, occupation\} \subset A_u$ and $genre \in A_i$. G refers to the set of all user groups. We also define two functions $users(g)$ and $desc(g)$ that return g's members and g's description, respectively. $|G| = 2^{|V|+|I|}$, which can be very large even with a few attribute values and items. For example, with $|I| = 5$, $|A| = 4$ and 5 values per attribute (i.e., $|V| = 20$), $|G|$ will be in order of 10^7.

2   One-Shot Exploration

We describe motivating examples and then summarize our contributions and takeaways for one-shot user data exploration.

Example 1 (Finding a Suitable Movie).   Sofia wants to see a comedy. IMDb gives her is a global average rating or an average rating broken down by pre-packaged user segments. Sofia would like to select a set of comedies and only wants to see segments where users have rated those comedies similarly. Those segments, akin to user groups, must be mined from the rating records of her selected comedies and cannot be pre-packaged for every possible subset of D.

Example 2 (Quantified-Self).   Mary[1] is an avid book reader and is very active on BOOKCROSSING. She has over 1,000 ratings (ranging from 1 to 10 but mostly high) for her favorite author, *Debbie Macomber*. She is looking to join an online book club where she can find people with whom she agrees and disagrees to engage in a stimulating debate. Returning user groups that highly agree or disagree with Mary would be useful to her.

Summary of Approaches. The first collection of papers state the problem of user data exploration as a one-shot optimization where the input is any subset of the database D and the output is k user groups in G. In [2], we stated the problem as finding k groups whose ratings are uniform or polarized. This could help us solve Sofia's problem in Example 1. Our follow up work [3], examined the case where the k groups are characterized by similar/dissimilar tagging actions. In [1], the input to the problem was stated as finding a set of groups whose rating distribution was close to one distribution provided as input. This could help us solve Mary's problem in Example 2. Finally, in [4], we studied a multi-objective formulation of the problem of finding user groups where criteria such as the coverage of input data, the rating distribution of individual groups, or the diversity of returned groups, could be optimized together.

Summary of Takeaways. In this series of papers, we showed that it is useful to state user data exploration as the problem of optimizing local and global criteria. Individual group size, the distribution of ratings within a group, or the length of a group's description are examples of local criteria. Global criteria refer to the coverage of input records or the diversity of returned groups. We showed the hardness of our problems and proposed heuristics to address them efficiently.

---

[1] *We do not know Mary personally but she is a real user on* BOOKCROSSING.

3   Interactive Exploration

We describe a motivating example and then summarize our contributions and takeaways for interactive user data exploration studied in [5].

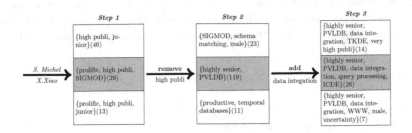

Fig. 1. WEBDB 2014 program committee formation

Example 3 (Expert-Set Formation). Martin[2] is looking to build a program committee formed by geographically distributed male and female researchers with different seniority and expertise levels. Martin starts with 2 junior researchers, S. Michel and X. Xiao (Fig. 1). The system uses them to discover 3 groups out of which the group described as *prolific, high publications* and publishing in *SIGMOD*, has 29 geographically-distributed and gender-distributed researchers. Martin chooses L. Popa, A. Doan, M. Benedikt, and S. Amer-Yahia in those groups. Step 3 reveals 3 other groups out of which 1 group contains 119 highly senior researchers and can be broken into 3 sub-groups whose researchers have expertise in *data integration*. In particular, the group labeled *query processing, PVLDB* and *ICDE* contains 26 senior researchers out of which 8 are of interest to the PC chairs: J. Wang, F. Bonchi, K. Chakrabarti, P. Fratenali, D. Barbosa, F. Naumann, Y. Velegrakis and X. Zhou. At this stage and after 3 steps only, Martin covered 80% of the WEBDB PC.

Summary of Approach. We developed a framework where, at each iteration, an analyst visualizes k groups, chooses one group of interest, and takes an action on that group (add/remove members, modify description). The analyst then chooses an operation to discover k diverse groups that are relevant to the current group. We provide two operations: exploit generates k diverse groups that cover the current group, and explore that generates k diverse groups that overlap with the current group. We showed that 50% of the program committees of conferences such as SIGMOD and VLDB and CIKM, can be built in fewer than 9 interactions. We also showed that 80% of the program committee of SIGMOD can be built in 10 steps and that it is closer to 15 for CIKM. The diversity of topics in CIKM increases the needed steps to cover more committee members.

---

[2] *Martin Theobald was indeed the WEBDB PC chair in 2014!*

Summary of Takeaways. In this work, we showed the usefulness of interactive group exploitation and exploration and formulated them as optimization problems. We proved their hardness and devised greedy algorithms to help analysts navigate in the space of groups and reach one or several target users. One important takeaway in interactive exploration is the need for a principled evaluation methodology. That requires the definition of objective measures such as the number of steps necessary to find a single user or a set of users as for Martin in Example 3. It also requires the careful design of appropriate user studies.

4   Visual Analytics

In many exploration scenarios, the analyst only has a partial *partial understanding of her needs* and needs to refine them as she extracts more insights from the data. Ideally, she would be immersed in an environment where she could provide any kind of input related to her needs be it a dataset, rating distributions or a query of interest. The proposed environment would take her input and start suggesting user groups along with some analytics. Such a tool would provide the ability to find and connect user groups in a way that seemlessly integrates our contributions described in Sects. 2 and 3. User groups could be visualized in a directed force layout to prevent visual clutter. Histograms and charts that show detailed statistics about groups could be provided. Those statistics must be displayed in coordinated views where a brush on one (e.g., histogram) updates all others instantaneously. In addition to exploit/explore primitives, there is a need to provide long jumps as well as the ability of undoing a previous step. An essential element of this tool would be to let analysts provide explicit feedback in addition to the implicit feedback gathered from their various actions. Feedback will help to customize the exploration by recommending subsequent exploration steps. It will also help evaluate the usefulness of such a tool.

References

1. Amer-Yahia, S., Kleisarchaki, S., Kolloju, N.K., Lakshmanan, L.V., Zamar, R.H.: Exploring rated datasets with rating maps. In: WWW, Perth, Australia (2017, to appear)
2. Das, M., Amer-Yahia, S., Das, G., Yu, C.: MRI: meaningful interpretations of collaborative ratings. PVLDB **4**(11), 1063–1074 (2011)
3. Das, M., Thirumuruganathan, S., Amer-Yahia, S., Das, G., Yu, C.: An expressive framework and efficient algorithms for the analysis of collaborative tagging. VLDB J. **23**(2), 201–226 (2014)
4. Omidvar-Tehrani, B., Amer-Yahia, S., Dutot, P.-F., Trystram, D.: Multi-objective group discovery on the social web. In: Frasconi, P., Landwehr, N., Manco, G., Vreeken, J. (eds.) ECML PKDD 2016. LNCS, vol. 9851, pp. 296–312. Springer, Cham (2016). doi:10.1007/978-3-319-46128-1_19
5. B. Omidvar-Tehrani, S. Amer-Yahia, and A. Termier. Interactive user group analysis. In CIKM, pp. 403–412. ACM, 2015

Data Wrangling and Data Integration

A Conceptual Approach to Traffic Data Wrangling

Mashael Aljubairah[1]([✉]), Sandra Sampaio[1], Hapsoro Adi Permana[1],
and Pedro Sampaio[2]

[1] School of Computer Science, University of Manchester, Manchester, UK
{mashael.al-jubairah,hapsoroadi.permana}@postgrad.manchester.ac.uk,
s.sampaio@manchester.ac.uk
[2] Alliance Manchester Business School, University of Manchester, Manchester, UK
p.sampaio@manchester.ac.uk

Abstract. Data Wrangling (DW) is the subject of growing interest given its potential to improve data quality. DW applies interactive and iterative data profiling, cleaning, transformation, integration and visualization operations to improve the quality of data. Several domain independent DW tools have been developed to tackle data quality issues across domains. Using generic data wrangling tools requires a time-consuming and costly DW process often involving advanced IT knowledge beyond the skills set of traffic analysts. In this paper, we propose a conceptual approach to data wrangling for traffic data by creating a domain-specific language for specifying traffic data wrangling tasks and an abstract set of wrangling operators that serve as the target conceptual construct for mapping domain-specific wrangling tasks. The conceptual approach discussed in this paper is tool-independent and platform agnostic and can be mapped into specific implementations of DW functions available in existing scripting languages and tools such as R, Python, Trifacta. Our aim is to enable a typical traffic analyst without expert Data Science knowledge to be able to perform basic DW tasks relevant to his domain.

Keywords: Data Wrangling · Data transformation and quality · Conceptual wrangling approaches

1 Introduction

Decision makers in different domains, such as healthcare, education and transportation, can gain significant advantages from the enormous volume of available data obtained from various data collection methods, such as site-based sensors, cell-phone tracking [1] and social media. However, data collected via these methods are prone to data quality problems, such as inaccuracy, incompleteness and heterogeneity [2,3]. More recently, data management techniques for data profiling, cleaning and integration have been adapted to improve the quality of large amounts of raw data, in preparation for analysis. The combination of

© Springer International Publishing AG 2017
A. Calì et al. (Eds.): BICOD 2017, LNCS 10365, pp. 9–22, 2017.
DOI: 10.1007/978-3-319-60795-5_2

these data management tasks is often called Data Wrangling (DW), generally defined as "the process by which the data required by an application is identified, extracted, cleaned and integrated, to yield a data set that is suitable for exploration and analysis" [4]. According to IBM [5], data analysts spend around 70% of their time conducting DW activities. Being an interactive and iterative process that involves the application of a variety of data management methods, and that generally lacks a rigid methodology across application domains, DW is often regarded as a highly complex job requiring advanced skills and domain expertise.

Data analysts typically perform DW tasks by using one or a combination of the two following approaches: (i) by programming their own DW applications, using languages such as Python, Java and R; and (ii) by interacting with existing DW tools, which often provide access via a Graphical User Interface (GUI). While approaches (i) and (ii) provide benefits, they also have shortcomings. Approach (i), for example, is often associated with completeness of functionality for fulfilling the requirements of the application in consideration; however, it also involves complex application development, advanced programming skills and brittle solutions that cannot be easily applied over data from other sources than the ones for which the solution was originally designed. On the other hand, approach (ii) is often associated with ease of user interaction, limited need for programming skills, provision of generic functionality that cannot be easily adapted to fulfil specific functional requirements, need for use of multiple tools to perform a single DW job, and limited opportunity for optimization.

To mitigate the limitations identified in approaches (i) and (ii), we propose a conceptual approach and an architectural solution for DW that combines advantages from (i) and (ii), while offering user interaction via a GUI and a high-level and domain-specific declarative language for simple DW tasks. The proposed architecture combines functionality from multiple DW tools and access to multiple data sources, by using Web Services technology. The result is an extensible DW tool, able to take advantage of DW functionality implemented within a variety of existing DW tools, and that provides extensibility and flexibility to allow data analysts to add functionality specific to the requirements of the application in consideration, creating a rich set of DW functions, that can be combined to accomplish simple and complex, general and domain specific DW tasks. For that, high-level user DW requests are automatically mapped into a set of conceptual DW constructs, that are ultimately translated into an execution plan represented as a workflow combining local as well as remote DW functions implemented across a multitude of tools. The proposed approach is tested with use cases from the Urban Traffic Domain, in which DW tasks associated with common data analysis requests by traffic analysts are executed over an implementation of the approach. The paper is organized as follows: Sect. 2 provides a literature review. Section 3 provides the conceptual DW approach including the architecture, conceptual and physical layer design sections, and discussion of implementation aspects. Section 4 provides the conclusions and future work.

2   Literature Review

There is a body of research to facilitate DW via Graphical User Interfaces (GUIs) and Domain-Specific Languages (DSLs). For example, in the work of *Kandel et al.* [6], the complexity of conducting DW was decreased, by associating DW functionality with data visualization constructs, allowing users to conduct DW via a visual interface. Tools offering DW functionality, such as Trifacta [7] and OpenRefine [8], provide concise DSLs combined with GUIs to isolate users from the complexities involved in the wrangling process. Even though both the Trifacta Wrangling Language and the General Refine Expression Language provide data transformation capabilities in Trifacta and OpenRefine, respectively, these languages are not high-level and declarative, and force users to use low-level language constructs. In addition, the level of completeness of functionality provided by these tools varies based on the DW requirements associated with the task at hand, as well as the characteristics of the target data. Considered in isolation, each of these tools will often fail to provide all the operations needed to effectively support the DW capabilities that complex information management problems typically require. A functionality-based comparison for the tools is provided in Tables 3 and 4 in Appendix A.

The difficulties in finding a tool that offers all the functionality required to perform DW tasks forces data analysts to face a steep learning curve before familiarising themselves with multiple tools and experiencing a rather laborious and complex process, in which data often needs to be transformed/reformatted to be transferred between different tools. In addition, data analysts may still have to use low-level programming constructs implemented in languages such as R, Python or Java to be able to customise code and solve specific data quality issues, despite using the DW tools. For example, Bluetooth-based road sensors used to collect traffic data often produce duplicate records for the same moving object due to multiple passengers carrying switched-on Bluetooth devices in a vehicle. For removing duplicates, in this case, multiple attributes need to be considered, such as vehicle identifier, time and location of detection, and device MAC address. However, tools offering DW functionality such as Trifacta Wrangler, OpenRefine and Talend data preparation only provide generic functionality for removing duplicates and so, cannot easily identify non-identical records as duplicates. In addition, because road sensors are generally prone to failure due to environmental conditions and are often able to detect only moving objects equipped with a switched on Bluetooth device, missing data is a common problem in traffic data sets. In the traffic domain, missing data can be replaced with data from the nearest periods or from similar locations or both [9]. Therefore, `Spatial Joins` using latitude and longitude information are often required to address missing data problems and are often not supported by general DW tools. Outliers are also difficult to address in generic DW tools due to the need to include semantic-based information to distingush between outliers and noise. Although there are outlier detection operations distributed across several DW tools, they do not correlate data with other attributes that are important to decide whether the value is an outlier.

3   A Conceptual Approach to Traffic Data Wrangling

3.1   Architectural Overview

Figure 1 illustrates the proposed Data Wrangling (DW) architecture. DW requests are expressed in the Declarative Data Wrangling Language (D^{2}WL) and submitted for parsing, during which validation of a D^{2}WL expression is carried out by checking the relevant data sources and other schema information, using meta data. At the end of parsing, a number of expressions in the Data Wrangling Language (DWL) is generated and submitted to optimization. In the current prototype, optimization is based on static information and heuristics, but future work will focus on adaptive and dynamic optimization, based on a cost model. The declarative language is designed for data analysts with limited or no programming skills and so it is based on a small number of clauses that define the location and format of input data sources (the FROM clause), the location and format of results (the TO clause), the main data wrangling activity to be carried out (the WRANGLE clause), and other data wrangling activities using clauses such as GROUPBY. A number of simple functions is also defined within D^{2}WL to facilitate the expression of specific data wrangling functions, such as filling in of missing values using function MISSINGDATA(listofattributes) and BY aggregate operation, which defines which data attributes should have their missing values filled in, and how. Figure 3 provides an example D^{2}WL expression that requests attributes gap and headway of a comma-separated values (CSV) file to be filled in by the average value for each of these attributes. Figure 2 provides the same DW request expressed in English. Note that gap (the time distance between the rear bumper of one vehicle and the front bumper of the vehicle behind it) and headway (the time distance between the front bumpers of two subsequent vehicles) are important traffic congestion indicators, because there is a direct correlation between gap or headway between two vehicles and vehicle speed, and an inverse correlation between the gap or headway and traffic volume. The design of the language is based on an analysis of the most common DW requests by traffic analysts and is able to express a majority of traffic DW requests. The architecture is designed to minimize human interaction by providing automatic translation of D^{2}WL expressions into DWL expressions, which are composed of a series of conceptual DW operators needed to perform the requested DW tasks. A GUI will also be provided, via which analysts are able to build complex DW requests, by connecting data sources and abstractions of DW operations as a workflow. Ultimately, each optimized DWL expression is mapped into a number of physical DW operations available as services and possibly from different tools and sources.

3.2   The Conceptual Layer

The model of data and computation of the proposed Data Wrangling Language (DWL) is functional, with a number of data types and functions, and

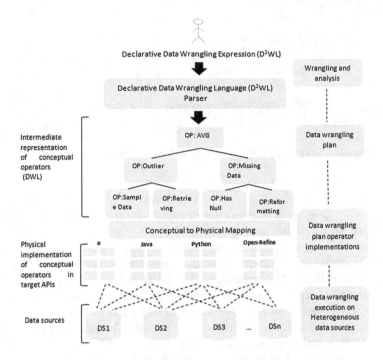

Fig. 1. Data wrangling system architecture.

"Fill in missing values using the average value for headway and gap, and based on observation site, vehicle direction, lane, hour, and day of week, the values for headway and gap could be inferred"

Fig. 2. English expression for the example DW task.

FROM   http://www.informationfortraffic.com/Inductive_loops.csv
TO   http://www.informationfortraffic.com/Inductive loops.csv
WRANGLE   MISSINGDATA(gap, headway)
BY   Average
GROUPBY   (siteID,vehicleDirection,lane,hour,DayOfWeek)

Fig. 3. D^2WL expression for the example DW task.

was designed to be extensible, allowing easy incorporation of additional functions. The data types supported are classified into two main types: atomic and aggregate. Atomic data types include string, character, date, time, boolean and numeric data, such as integer and float. While aggregate data types include general collection types, such as bag, representing an unordered collection of objects allowing duplicates, list, an ordered collection of objects allowing duplicates, set, an unordered collection of objects with no duplicates, and tuple, a record containing objects with different data types. The following symbols are relevant to

$$Result \Leftarrow view$$
$$(fill_{<<listofexpression>>,<<listofcondition>>}$$
$$(groupBy_{<<listofcolumnname>>}$$
$$(selectColumn_{<<listofcolumnname>>}$$
$$(addColumn_{<<listofexpression>>}$$
$$(enrichTimeStamp_{columnname,format,startdate}$$
$$(read(URL)))))))$$

Fig. 4. Conceptual DWL expression for the example DW task.

the DWL example described in Fig. 4: The symbol ⌐ is used to represent an empty bag, $<<>>$ is used to denote an empty list, [] represents an empty set and {} is used to represent an empty tuple. The Collection data type is a generalization of all four collection types, and its constructor is represented by (). An additional data type is Graph, which models visual representations of data.

Figure 4 describes a DWL expression for the D^2WL expression in Fig. 3. The **read** operator takes an URL as input identifying the name, format and location of the input file, possibly on a remote server, and brings the file into context. In this example, this CSV file is described in Appendix A. The **enrichTimeStamp** operator takes the **columnname** of a date attribute, and formats it according to the provided **format** and **startdate** to add date, month and year information to the attribute. The **addColumn** operator takes a **listofexpressions** containing pairs of the type **<<columnname, expression>>**, describing the names of new columns to be inserted into the file and expressions associated with the column names defining how each new column is to be populated. In this example, the pairs are as follows: **<<hour, hour=extractHour(completetimestamp)>**, **<dayofweek, dayofweek=extractDayofweek(completetimestamp)>>**.
extract Hour is used to extract the hour from **completetimestamp** which is the attribute generated by **enrichTimeStamp**. **extractDayofweek** is similar to **extractHour** but used to extract the weekday from a date. The result of operator **addColumn** is the input to **selectColumn** with a **listofcolumnnames**, representing the attributes to be projected from the file. The result of **selectColumn** is input to **groupBy**, which is also provided with a **listofcolumnames** that represent the attributes by which records are to be grouped. The **fill** operator then takes the result of **groupBy** and replaces the zero values in columns gap and headway with the value resulting from the evaluation of the expression provided in **listofexpressions**. In the example, the expressions in the list are **<< gap=average(gap), headway=average(headway)>>**, based on the **listofconditions** which are **<<if gap==0, if headway==0>>**. Note that the aggregate operations defined in the list of expressions are impacted by how records are grouped by a **groupBy** operator. Finally, the **view** operator allows the user to visualize the result on the screen. A further description of each operation is provided in Appendix B.

3.3   The Physical Layer

The Taverna [10] workflow management system is currently being used to call and manage several DW operations implemented within different target tools. The rationale behind the choice of Taverna in the current implementation of the proposed system is its ability to manage complex workflows that require connections to remote services, as well as its ability to allow the annotation and storage of a potentially large number of workflows. Figure 5 shows an example of how a conceptual DW operation such as **read** (described in Table 5 of Appendix B) is mapped into a low-level Taverna workflow. Each DW operation is represented as a Taverna workflow, so that the conceptual DW operation name is mapped to a workflow name, the operation argument(s) are mapped into the workflow input port(s) and the operation result(s) are mapped into the workflow output port(s). Each Taverna workflow of a mapped conceptual DW operation includes three processors each of which is associated with a Beanshell or REST service to specify how its functionality is achieved. These services and processors process and pass the input to the workflow via data links which have sink and source variables to identify the connection direction and obtain the target result. Each input and output to a processor is associated with the required input and output to its task-related activity, and the corresponding data link is used to associate the output of the processor with the input to the next processor. Figure 6 shows the actual execution steps of the Taverna workflow representing the **read** operation which is implemented as an R web service. The Beanshell service, **EncodeURLfile**, is responsible for encoding the input to the REST

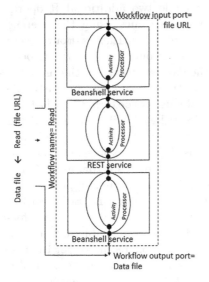

Fig. 5. Details of the Taverna workflow for the conceptual read operation.

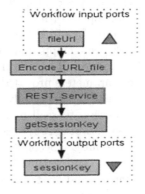

Fig. 6. Execution steps of a Taverna workflow representing the conceptual read operation.

service, `read.csv`, implemented in an R server. This is followed by the Beanshell service, `getSessionKey`, which obtains the session key associated with the data set brought into context by the `read.csv` function and makes it available to the next DW conceptual operator, another workflow that will process the data set. In Sect. 3.4, further implementation details for the example DW task in Fig. 4 are provided.

3.4   Implementation of DW Operations in Taverna

Figure 7 shows the realization of the DW task shown in Fig. 4 as a composition of Taverna workflows. Note that the DWL operators in Fig. 4 have been mapped into one or more low-level functions available from two remote servers, one being an R server, and the other one offering DW functionality via simple Python scripts. As described in Sect. 3.3, Beanshell services have been used to facilitate the HTTP-based communication of data between the various workflows, such as the URL of the file to be input to the task, and a session key associated with an intermediate result that is passed from one workflow to the next. Therefore, each workflow is, in turn, composed of three smaller services, specifically one REST service that encapsulates the main data processing activity associated with the workflow, and two enveloping Beanshell services, the first serving as an input data facilitator and the other one dedicated to extracting the data session key to be output to the next workflow, as Fig. 5 suggests.

In the workflow composition in Fig. 7 the DWL operator, `fill`, has been mapped into the following calls to R functions: `mutate`, to create new columns and impute missing values for `gap` and `headway` using expressions and column names as parameters. The expressions include two additional R operators, `replace` and `average`, to replace the missing values using the average value. Two Beanshell services, `imputeexpression` and `colnamesimpute` were used to provide the `mutate` parameters values; the `groupBy` operator has been mapped into R's `groupby` operator to group the data set based on the provided attributes given by a Beanshell service named `colnamesgroupby`. Similarly, the `selectColumn` operator has been mapped into R's `selectcolumns` to select the required columns provided by `colnamesslect` Beanshell service; the `addColumn` operator has been mapped into R's `mutate` function to create new columns contain information of hour and day of week based on the expressions and column names parameters provided by Beanshell services `expressionmutate2` and `colnamesmutate2`, respectively. The `read` operator has been mapped into R `read.csv` function to read a remote CSV file, and the `enrichTimeStamp` has been mapped into `completetimestamp` Python function to convert the time format from <Minute><Second><Milliseconds> to a more readable format (i.e. 24-hour format) and calculate the date. This operation required `begindate`, `datecolumnname` and `trafficdataurl` as parameters. These parameters are also the workflow input ports as shown in Fig. 7. In addition, Taverna allows downloading of the file using the REST service component named `getcsv`, and the data set was stored into a local CSV file using the Taverna input-output component named `writetextfile`. The different colours used in Fig. 7 indicate

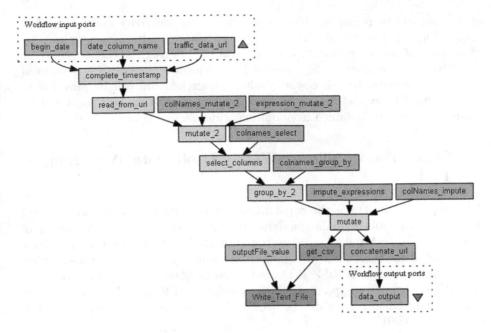

Fig. 7. Implementation of DW task using Taverna.

different services provided by Taverna. The pink rectangles indicate the encapsulated individual workflows, the brown rectangles indicate Beanshell services, the blue rectangles indicate REST service and the purple rectangle indicates Taverna input-output component.

4   Conclusions and Future Work

Due to the complexity associated with the DW process, there is often the need to apply several DW tools, presenting significant challenges to the data analyst. To address these challenges, a novel architectural approach to DW has been proposed. The proposed architecture for DW combines advantages from the existing DW approaches by providing abstract and domain-specific DW constructs without the need for end users to learn low-level programming APIs. In addition, this approach allows data analysts to take advantage of the functionalities available in existing tools. In addition to the Declarative Data Wrangling Language, D^2WL, the conceptual Data Wrangling Language (DWL) can also be complemented by other high-level and declarative domain-specific languages relating to other domains such as education, healthcare and finance. We are currently developing a graphical user interface for traffic analysts to specify D^2WL expressions and also assistive technology to map from high-level requests to optimised DW strategies implemented in the target platforms. To evaluate the proposed tool, a number of end user experiments will be conducted where analysts from traffic domain are to conduct data wrangling using the tool. A detailed evaluation of the tool will be performed based on both user productivity and tool performance. The user productivity will be measured based on criteria such as learning curve, ease of use, functionality coverage and level of user interaction/effort by quantifying the number of steps required to perform a number of DW tasks of varying complexity. To collect the results, a questionnaire will be provided to data wranglers (i.e. end-users) and performance of existing DW tools will be compared to results obtained from our experiments.

A   Comparison of Three Widely Used Data Wrangling Tools

This Appendix illustrates the capabilities and limitations of three widely used data wrangling tools in fulfilling a data wrangling task using the traffic data sets depicted in Tables 1 and 2, and illustrated in Table 3. In addition, it provides a comparison of the three data wrangling tools according to the import and export data format as shown in Table 4. The data wrangling task described as "Retrieve the value of the average speed of vehicles passing road R, (e.g. A572 Manchester Road, Astley Green), on a day of week D (e.g. Monday) at a time interval T (e.g. 8:00 to 9:00)."

B   Description of Data Wrangling Operations

This Appendix provides a description of data wrangling operations as shown in Tables 5, 6, 7, 8, 9, 10 and 11.

Table 1. Data collected by inductive loops, taken from TfGM database [11]

Site ID	Date	Lane	Lane Name	Direction	Direction Name	Class Scheme	Class	Class Name	Length	Headway	Gap	Speed (mph)	Weight	Vehicle Id	Flags	Flag Text	Num Axles
'000000001304	00:00:29	1	EB	1	East	3	2	Car	14.1			33.6		0			0
'000000001304	00:01:30	1	EB	1	East	3	2	Car	14.4	60.8		60.5	30.4	0			0

Table 2. Static data, taken from TfGM database [11]

Site ID	Site Name	Description	Speed Limit (mph)	Grid	Orientation	Longitude	Latitude	Bearing	Parameters
1304	1304	A572 Manchester Road, Astley Green	30	369151000000	E	-2.46646	53.49416	90	noexport=1
1305	1305	A572 Leigh Road, Boothstown	40	373097000000	w	-2.40707	53.50274	270	noexport=1

Table 3. Applying the data wrangling task on data wrangling tools

Stage No	Description	OpenRefine	Trifacta-free desktop version	Talend-data preparation tool
Stage 1	Check the datasets characteristics:big data	Not supported	Supported	Not supported
Stage 2	Check the datasets characteristics:CSV data format	Supported	Supported	Supported
Stage 3	Enrich the dataset by Time and Date	Not supported	Not supported	Not supported
Stage 4	Extract the last four digits of Site ID	Supported	Supported	Supported
Stage 5	Extract weekday	Supported	Supported	Supported
Stage 6	Join the data with static data using SiteID as a key	Supported	Supported	Supported
Stage 7	Filter the data based on weekday, road name and time	Supported	Supported	Supported
Stage 8	Calculate the average	Not supported	Supported	Not supported

Table 4. Data format-based comparison of the data wrangling tools

Comparison criteria	Trifacta-free desktop version	OpenRefine	Talend-data preparation tool
Input data format			
Comma Separated Values (CSV)	Supported	Supported	Supported
JavaScript Object Notation(JSON)	Not supported	Supported	Not supported
Extensible Markup Language (XML)	Not supported	Supported	Not supported
Export data format			
Comma Separated Values (CSV)	Supported	Supported	Supported
JavaScript Object Notation(JSON)	Not supported	Not supported	Not supported
Extensible Markup Language (XML)	Not supported	Not supported	Not supported

Table 5. Operation 1

Operation name	read
Format/Operation expression	read(String)
Arguments/Inputs	String: the string includes the file name, type and its location on a remote server which can be represented in a URL
Description	"read" reads a file in a remote server by specifying the name, type and its location using a URL
Output	Collection data type

Table 6. Operation 2

Operation name	enrichTimeStamp
Format/Operation expression	enrichTimeStamp (({Literal}), String, Time, Date)
Arguments/Inputs	({Literal}): the data set where you want to perform the "enrichTimeStamp" operation. The collection can be a bag, set or list of tuple data type. String: the column name in your data set which includes Time data that need to be converted. Time: the specified Time format provided by a user Date: the start date specified by a user
Description	"enrichTimeStamp" enriches a data set by converting the time format from MM:SS:MS to time HH:MM:SS and infer date as DD/MM/YYYY. This would be based on the provided Time format and start date
Output	Collection data type

Table 7. Operation 3

Operation name	addColumn
Format/Operation expression	addColumn (({Literal}), << Expression >>)
Arguments/Inputs	({Literal}): the data set where you want to perform the "addColumn" operation. The collection can be a bag, set or list of tuple data type. << Expression >>: list of expressions contains the name/s of new column/s and how the data will be filled in the new column/s The expression can be: (1) comparison expressions: results in a value of either TRUE or FALSE. The expression can include relational operators and operators such as AND, OR, XOR, NOR, and NOT. (2) Arithmetic expression: results in a numeric value. The expression can include arithmetic operators (e.g. $+$, $-$, $*$, $/$, $\%$) The expression can also include one or more operations
Description	"addColumn" enriches the data set by adding new column/s filled based on the specification written in the expression
Output	Collection data type

Table 8. Operation 4

Operation name	selectColumn
Format/Operation expression	selectColumn (({Literal}), << String >>)
Arguments/Inputs	({Literal}): the data set where you want to perform the "selectColumn" operation. The collection can be a bag, set or list of tuple data type. << String >>: list of column names represents the selected columns
Description	"selectColumn" selects specific column/s from a data set
Output	Collection data type

Table 9. Operation 5

Operation name	groupBy
Format/Operation expression	groupBy (({Literal}), << String >>)
Arguments/Inputs	({Literal}): the data set where you want to perform the "groupBy" operation. The collection can be a bag, set or list of tuple data type. << String >>: list of column name/s
Description	"groupBy" arranges identical data into groups based on the specified columns
Output	Collection data type

Table 10. Operation 6

Operation name	fill
Format/Operation expression	fill ({Literal}), << Expression >>, << Expression >>)
Arguments/Inputs	({Literal}): the data set where you want to perform the "fill" operation. The collection can be a bag, set or list of tuple data type. << Expression >>: list of expressions contains the name/s of columns which have missing values and how the values will be resolved./s. The expression can also include one or more operations. << Expression >>: list of conditional expressions contains the name/s of columns which have missing values and conditions on the values of these columns to identify when the value is considered as missing and should be replaced. The expression is comparison expressions which result in a value of either TRUE or FALSE. The expression can include relational operators and operators such as AND, OR, XOR, NOR, and NOT
Description	"fill" replaces every row in specified column/s by the given value based on the conditional expression provided by a user
Output	Collection data type

Table 11. Operation 7

Operation name	view
Format/Operation expression	view ((\{Literal\}))
Arguments/Inputs	(\{Literal\}): the data set where you want to perform the "view" operation. The collection can be a bag, set or list of tuple data type
Description	"view" allows users to visualize a file on a screen

References

1. Lopes, J., Bento, J., Huang, E., Antoniou, C., Ben-Akiva, M.: Traffic and mobility data collection for real-time applications. In: 13th International IEEE Conference on Intelligent Transportation Systems (ITSC), pp. 216–223. IEEE (2010)
2. Hutchins, J., Ihler, A., Smyth, P.: Probabilistic analysis of a large-scale urban traffic sensor data set. In: Gaber, M.M., Vatsavai, R.R., Omitaomu, O.A., Gama, J., Chawla, N.V., Ganguly, A.R. (eds.) Sensor-KDD 2008. LNCS, vol. 5840, pp. 94–114. Springer, Heidelberg (2010). doi:10.1007/978-3-642-12519-5_6
3. Jagadish, H.V., Gehrke, J., Labrinidis, A., Papakonstantinou, Y., Patel, J.M., Ramakrishnan, R., Shahabi, C.: Big data and its technical challenges. Commun. ACM **57**(7), 86–94 (2014)
4. Furche, T., Gottlob, G., Libkin, L., Orsi, G., Paton, N.W.: Data wrangling for big data: Challenges and opportunities. In: EDBT, pp. 473–478 (2016)
5. Terrizzano, I., Schwarz, P.M., Roth, M., Colino, J.E.: Data wrangling: the challenging journey from the wild to the lake. In: CIDR (2015)
6. Kandel, S., Heer, J., Plaisant, C., Kennedy, J., van Ham, F., Riche, N.H., Weaver, C., Lee, B., Brodbeck, D., Buono, P.: Research directions in data wrangling: visualizations and transformations for usable and credible data. Inform. Vis. **10**(4), 271–288 (2011)
7. Trifacta Wrangler Enterprise. Trifacta wrangler (2015). https://www.trifacta.com/. Accessed 05 Jan 2016
8. Open source community Google. Openrefine tool (2015). https://github.com/OpenRefine/OpenRefine/wiki/General-Refine-Expression-Language. Accessed 28 Dec 2015
9. Guo, C., Jensen, C.S., Yang, B.: Towards total traffic awareness. ACM SIGMOD Rec. **43**(3), 18–23 (2014)
10. Apache Software Foundation. Taverna (2015). https://taverna.incubator.apache.org/. Accessed 30 Jan 2017
11. Transport for Great Manchester (2015). http://www.tfgm.com/Pages/default.aspx. Accessed 17 Jun 2015

Integrating Online Data for Smart City Data Marts

Michael Scriney[1](✉), Martin F. O'Connor[2], and Mark Roantree[1]

[1] Insight Centre for Data Analytics, School of Computing, Dublin City University,
Glasnevin, Dublin 9, Ireland
`michael.scriney@insight-centre.org, mark.roantree@dcu.ie`
[2] Department of Computing, Institute of Technology Tallaght, Dublin 24, Ireland
`martin.oconnor@it-tallaght.ie`

Abstract. The development of smart city infrastructures is seen as an important strategic component for many countries with applications in the areas of government services, healthcare, transport and traffic management, energy, water, and the management of waste services. In general, each of these infrastructural components generates data as it delivers its service. The usage of this data is often crucial both to the continued development of the service and for forward planning. While the data sources are often not complex in structure, traditional decision support systems are not built to handle continuous data streams, new or disappearing data sources, or the integration of multiple online data sources. In this work, we develop a flexible ETL process to quickly process and integrate online smart city data sources to deliver information that can be used in close to real-time.

1 Introduction

The Internet of Things has provided a means of collecting and analysing previously inaccessible data across a wide variety of topics. Recently, there has been an emergence of publicly available data regarding the infrastructure of various cities. This data, provided by governmental agencies and private industry can lead to direct benefits for the citizens of a city. The projects which process, store and analyse this data are known as Smart City applications. Data in a smart city is generated from a wide variety of sources across a large number of domains such as housing, environment, transport and so on. Similar to traditional web data, this data is available on-line, typically in either XML, JSON or CSV format and there have been many approaches to building smart city applications [9] and to clustering or integrating query graphs on smart city data [13]. Most of the query or search engines are RDF based eg. [4] but recently, we have seen the emergence of approaches to understanding JSON schemas [1] which allows for scaling of large online datasets.

This work is supported by Science Foundation Ireland under grant number SFI/12/RC/2289.

© Springer International Publishing AG 2017
A. Calì et al. (Eds.): BICOD 2017, LNCS 10365, pp. 23–35, 2017.
DOI: 10.1007/978-3-319-60795-5_3

Decision makers who use smart city applications expect to have some form of OLAP service which provides the datasets from a warehouse upon which they make their decision. However, online data sources are not included in many data warehouses for many reasons. These may include lack of control of the source data, lack of understanding of how to process the data, a constantly changing structure to the data or issues to do with quality control in terms of the data's content. Access to these services generally requires an API supplied by the service provider or some form of wrapper to extract the data. In either case, there is a time lapse between when new online information becomes available and when it is accessible using OLAP tools for decision makers. Consider a scenario where emergency services require the fastest route between two points in the city: unless data is available in close to real-time it is of little use to the route planning service. In this paper, we will present an approach to making online data available in smart city data marts in close to real time. To deliver this service, we update a Data Lake every 30 s from source data. Briefly, a Data Lake is a data store which stores data in its original native format. By close to real time, we mean that online information providers are checked every 30 s for updates, those updates (if any) are populated into data marts that use our Star Graph format [15] to provide data immediately to decision makers.

1.1 Background and Problem Statement

In order to deliver integrated datasets from multiple external sources, what is required is a sufficiently robust Extract-Transform-Load (ETL) process so that facts and dimensions can easily be identified; external sources are mapped to some form of ontology, and an integration process to merge the specific data sources that are needed for specific tasks. An approach advocated in [2] is to use lightweight dynamic semantics which our system uses to maintain interoperable descriptions of data sources inside the Data Lake. This metadata is then used to update a data mart in close to real-time. In our approach, we do not assume that a predefined schema and ETL process exists. This is traditional ETL where the warehouse schema is designed upfront, and data is extracted from all (known) sources in advance, integrated according to strict rules and data is loaded whether it is required or not.

The problem with online data sources is that they evolve much more than in-house databases. Sources can disappear or change structure; new sources can become available; different user needs often require the construction of heterogeneous data marts. The problem with applications such as smart cities are that in order to exploit the wealth of information available, it requires a more flexible approach to ETL and data mart construction. At the same time, we aim to provide OLAP style functionality to create the datasets necessary for analysis of long duration data or small data marts with only the most recent updates. This solution must also take into consideration that data marts should be dropped where data sources are no longer available or newly created where new sources have become available as data marts may use different combinations of source data.

Contribution. In earlier work [15], we presented a system to automatically create multi-dimensional data marts from online sources. We employ a graph representation for data marts which is called a *StarGraph*, a graph structure which comprises the multi-dimensional concepts of *facts* and *dimensions*. In this work, we present a mechanism for integrating StarGraphs. While the StarGraph has a 1-1 relationship with external sources, a second structure, the *ConstellationGraph* is an integration of StarGraphs, with 1-many mappings with the StarGraphs which comprise it. In addition, we can operate real-time data marts where data is pulled from the Data Lake containing online sources on demand.

Paper Structure. The paper is structured as follows: in Sect. 2, we present related research; in Sect. 3, our case study is introduced; in Sect. 4, we describe the processes involved in our system; in Sect. 5, we provide an evaluation of our work; and finally, in Sect. 6, we provide some conclusions.

2   Related Research

Similar to our research, the authors in [16] use graph structures to model heterogeneous data sources which are to be integrated and stored in a data warehouse. However, the authors employ the use of application designers to manually add an ontology to each graph structure in order to produce mappings from source to target schemas, whereas our process seeks to automatically identity facts, dimensions and measures and provide a corresponding mappings file for further analysis of data sources.

In [12], the authors present a means of providing multidimensional analysis through the use of a multidimensional ontology. This ontology itself is built from a semantic data warehouse, which provides ontologies about all source data required for multidimensional analysis. However, this does not provide the flexible and more lightweight approach [2] necessary for online data sources. In [3], the authors again present a smart city ontology which is combined with a data collection system to harvest smart city data sources to provide smart city applications. Similarly to our paper, the authors focus on transport data. However, while their system utilises an ontology in order to provide mappings and integration, our system is more flexible in that it automatically generates mapping files and identifies facts and dimensions for further analysis.

In [11] the authors present a real time smart city application which tracks and monitors city facility utilisation. The authors use a system which provides OLAP functionality on a data stream. However, the integration methodology for various data sources used by the authors is a manual process, which requires a custom wrapper to be developed per data source, whereas our system provides both a manual and automatic means of integrating data.

In [8] the authors detail a smart city system in Barcelona. The system is composed of a number of sensors available through HTTP/REST interfaces. The authors note there is a difficulty in the management of the data due to the large volume generated by multiple sensors. The smart-city sensors generate data in the form of JSON which is stored in a database made available to third-parties over XML. Similar to our integration approach, the authors note that an intelligence module is provided which correlates data in space and time, the authors do not however detail how this system works, neither do they detail how the data is integrated and stored in the central database.

3   Case Study Introduction

We will now introduce our case study which will be used to conceptualise the need for a StarGraph and provide a means of illustrating the processes involved throughout the lifecycle of our system. Commuters use various means of transport in order to reach their place of work. However, due to the ever increasing number of people who commute the transport infrastructure can become strained. This leads to increased commuting times. There are benefits to both the transport providers and the commuters themselves for minimising travel times across the city.

The transport infrastructure of a city can be conceptually viewed as a graph. A node on the graph indicates a point of departure or a destination (for example, a train station, a bus stop, a road junction and so on). An edge between two nodes indicates a path. A contiguous sequence of two or more paths is a route. This topological graph may be used to find a path for the commuter between their departure and destination with the shortest distance. However, when commuting, distance covered is not the only feature of interest. For many commuters, the minimum time taken to commute is preferable to distance covered.

In order to determine the most efficient route at time t a collection of historical transport data must be collected. To satisfy this case study, the following functionality must be provided:

1. Harvesting of multiple domain-related data sources.
2. Integration and contextual enrichment of the data.
3. Population of the integrated data warehouse.
4. Provision of a Query Service.

A number of Ireland's transport providers publish real-time data, available through the use of public APIs. For the purposes of this case study the following datasets have been chosen.

- Dublin bus RTPI [6], details the next buses due at a specified bus stop.
- Dublin Bikes [5], provides real time indication of bike availability.
- LUAS (Tram line) [17], Details the next arrivals along a tram line.
- Motorway travel times [18], The travel time taken to traverse the motorways.
- Dublin City Travel Times [7], Time taken to traverse junctions on Dublin Roads.
- Irish Rail API [10], Details the upcoming arrivals at a given train station.

In this paper we will demonstrate how our system can be used to find the most time-effective means of getting a commuter to their destination.

4   Process Framework

There are four main processes to our system as shown in Fig. 1.

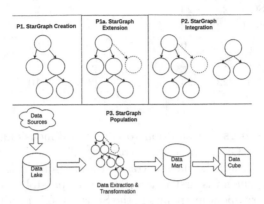

Fig. 1. A StarGraph-based ETL System

1. **P1. StarGraph Creation**. The first process in our system creates a Star-Graph from the source schema.
2. **P1a. StarGraph Transformation**. This is an optional step which can be used to enhance the structure of the StarGraph by allowing a user to manually edit the graph, or add supplementary information necessary for further analysis.
3. **P2. StarGraph integration**. This process takes one of two forms: integration of 2 StarGraphs to form a ConstellationGraph; or the integration of a StarGraph and ConstellationGraph which allows continued growth of the final schema.
4. **P3. StarGraph Population**. This process populates a fact table based on the mapping rules. It has two inputs: the mapping rules corresponding to the data mart (P3) and the data stream from which the data is to be extracted.

4.1   StarGraph Creation (P1)

A StarGraph represents the facts, dimensions and measures found inside a data source. In addition, it provides a mappings file which can be used to transform the source data into the StarGraph representation. For the case study presented here, when the last source file is updated, the StarGraph is then updated from all sources. For the 6-source example shown in Table 1, updates occur between 30 and 300 s.

Table 1 details the StarGraphs created from the data sources detailed in Sect. 3 along with the time taken to populate from the data streams and the number of new instances constructed for 1 sample update.

Table 1. Constructed StarGraphs

Name	ID	Format	Nodes	Edges	Dimensions	Measures	Population(ms)	Rows
Dublin Bikes	bik	JSON	14	14	2	4	69	245
Dublin Bus	bus	JSON	23	23	2	0	24	101
LUAS	lus	HTML	3	3	1	0	7	2
City Roads	rds	CSV	7	0	1	7	64	902
Motorway	mot	JSON	18	2	10	3	14	25
Irish Rail	ril	XML	22	23	1	2	10	2

Generating Mappings. In addition to the multidimensional representation produced by the StarGraph process, a corresponding series of mapping rules are produced to facilitate the data extraction and transformation from source to target. When a schema is first read, a list of all possible paths through the schema and their corresponding points in the StarGraph is produced. This metadata is used as a basis when formulating the Mapping Rules for the StarGraph. Mappings are stored in JSON format.

Mappings are only generated once the fact table has been constructed from the StarGraph. The mappings database contains:

- Dimension - Indicates a full dimension object.
- name - Indicates the name of either an attribute or measure.
- src -This attribute details the location of the relevant data in the data stream. It can take the form of an xpath query, JSON dot notation or a function definition (Sect. 4.2).
- id - Id denotes the primary key of a table. The id attribute can only be found inside an Dimension object or Subdimension. This can either be automatically generated (in the absence of a defined key) or is created using a src query.
- table - The name of the table to store the data
- atts - Atts is a list of attributes associated with a dimension or subdimension
- type - Type is the data type of the attribute.
- subdimensions - This is a list of subdimension objects.
- measures - These are a list of measures which are found.

4.2   Enrichment and Transformation (P1a)

Data Transformation. For some datasets, the measure may not immediately be detected. This issue can arise when a transformation function is necessary to expose the measure. For example, the Dublin Bus dataset does not provide a unique measure (i.e. when the next bus is due) but rather provides two timestamps one named `scheduledarrivaldatetime` which refers to the time the bus is expected at a stop, and another named `timestamp` which refers to the current timestamp. A useful measure would be when the next bus is due in terms of seconds (or minutes). With a created StarGraph and mappings file,

```
{"name":"timeInSeconds",
 "src":function(){ return
 $scheduledarrivaltime.getTime()- $timestamp.getTime(); },
 "type":"int"}
```

Fig. 2. Example of user defined transformation

this file can be extended by an application developer to provide a transformation function (Fig. 2).

By placing this function under the `Measures` section of the mappings file, the code will be executed. This result will then be stored in the fact table under the name `timeInSeconds`. If it is placed in the `Dimension` or `Subdimension` areas of the mappings file, it will be stored as an attribute of the Dimension.

At present, the mapping files are stored as JSON, and transformation functions are written in JavaScript. When new data is introduced to the system, the mappings file extracts all the source data for population. In addition, these transformation functions are evaluated in order to produce new measures for analysis as the system is being populated.

Providing Streaming Context. Many smart city data streams provide data in a format coded by an application designer, and not by a business user. An example of data coding would be providing unique id's to objects. However, oftentimes supplementary data relevant to the code is not provided as part of the data stream, but as a static file. These static files can contain geolocation data, fully resolved addresses etc. which are tied to the unique ids provided by the stream. Static file integration seeks to resolve these differences by re-combining the static supplementary data to the data stream.

An example of static file integration can be seen if we examine the City Roads dataset. The City Roads dataset is a CSV file which provides little information about the data being accessed (Fig. 3). This is very similar to body sensor applications which produce very high volumes of information [14] but where the data generated is very simplistic and requires external semantics to increase its impact.

#Route	Link	Direction	STT	AccSTT	TCS1	TCS2
1	1	1	128	128	2127	175

Fig. 3. Example of the City Roads dataset

In this data, the headers `TCS1` and `TCS2` refer to real world locations. However, without the supplementary descriptive data, the numeric data lacks meaning. The data provider (Dublin City Council) also provides a KML file (geographic annotation) (Fig. 4) which provides the additional data necessary to understand the real time data. This supplementary data can be used to enrich the existing

```
<SimpleData name="TCS1">6006</SimpleData>
<SimpleData name="TCS2">2031</SimpleData>
      <coordinates>-6.172557196923532,53.291529410712464
  -6.184390631942438,53.296472990543961</coordinates>
```

Fig. 4. KML file (truncated) provided by Dublin City Council

CSV file with latitudes and longitudes linking the data TCS1 and TCS2. The supplementary data is supplied by the same provider as the stream, as such there is a one-to-one mapping between the terminology used for the stream and the static file, therefore, this data can be integrated using a simple text matching operation which can be used to link the `coordinates` tag to the existing data. Additionally the mappings file associated with the StarGraph has been extended to map the data stream to the provided context.

4.3   StarGraph Integration (P2)

Two or more StarGraphs are integrated to produce a `Constellation Graph` with a process that can be either manual or automatic. For manual integration, a user selects an integration attribute for both StarGraphs and they are joined on that attribute. Once the StarGraphs have been joined, changes to dimensions and measures require the generation of new mappings. These mappings are similar to the mappings described in Sect. 4.1. They are generated in a similar fashion to that of a StarGraph, except in this instance an attribute may have multiple `src` properties. Automatic integration combines StarGraphs using an ontology and matching attributes based on name, location or date. For example Figs. 5 and 6 represent two StarGraphs which are to be merged. The nodes `datetime` from the Bikes StarGraph and `timestamp` from the Bus StarGraph are merged because they occupy the same date dimension (Fig. 7).

Fig. 5. Bikes StarGraph

Fig. 6. Dublin Bus StarGraph

4.4   Populating and Updating (P3)

As indicated earlier, the trigger for updating a ConstellationGraph is when the last source file has been updated. The mappings database is used to transfer data from the Data Lake into the warehouse (StarGraph).

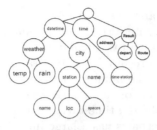

Fig. 7. ConstellationGraph constructed from the bikes and bus StarGraphs

Querying. Once a StarGraph (or ConstellationGraph) has been populated, the location data found inside the StarGraph can be used to construct a topology representing the physical distribution of the data. If there are multiple latitudes and longitudes occupying the same instance, an edge is created between them.

Recall the inputs for the case study are a points of departure and destinations and the datetime of departure. The first step in completing any valid query is to examine the topological graph to determine all possible paths which lead from the departure point to the destination. For the purposes of this case study this is achieved using a depth first search to determine all simple paths. Once all simple paths have been constructed, these are stored as a series of possible routes.

The next step is to examine which route is the fastest for the departure time the user provided. The historical aggregate of the travel time for each route at time t is evaluated and the minimum route is returned.

5   Evaluation

The goal of our evaluation was to measure how close to real-time, we can deliver the datasets necessary to compute information for travel requirements. The travel case study has six data sources and we used 57 different ConstellationGraph (schema) configurations to measure the usability of the system and potential scalability. Table 2 reports from a representative batch of 12 of these configurations. In each case, we report the time taken in milliseconds to refresh each schema when an update is detected. An update in this case is where change was detected in *all* data sources with the trigger occurring when the final data source has changed. Our experiments were completed using Node v6.10.0 64-bit, Mongo v3.4.2 64-bit on a SONY VAIO with 12 GB RAM and Intel i7-3632QM 2.20 GHz, Windows 10 Home edition 64-bit

Datasets were separated into large and small, based on the number of new instances, with **bus**, **bik** and **roads** constituting the large dataset, and the remainder making up the small one. We are not loading large data volumes as updates occur no longer than 5 min apart: the goal is to make current data available.

Column 2a was created from combining two sources from the large datasets (specifically, the **bus** and **bik** datasets). 2b was created from combining one large and one small dataset (bus and lus). This update generated 270 new instances and was completed in 76 ms. Interestingly, for population of two data sources, 2c which was created by combining two small data sources (lus and ril) took the largest time of 157 ms which resulted in a fact table consisting of only 4 rows. This indicates that it is not the number of rows generated that impacts more on time but rather the format of the data sources itself. As these sources are HTML and XML respectively, the time taken to parse these schemas compared their JSON and CSV counterparts has a large impact on the integration times.

Columns 3a, 3b and 3c show the times taken to populate a schema (data mart) composed of three sources. 3a was created from all three of the large data sources (bus, bik and rds), 3b was created from two large and one small data source (bus, bik and mot) and 3c was created from all of the small data sources (lus, mot, ril). Once again, the time taken to populate a fact table composed of all the large data sources, and the small ones were comparable (156 and 159 ms respectively). These times were despite the fact that the large data sources result have 1219 more instances than small sources.

Columns 4a, 4b and 4c show the times taken to construct a schema from four data sources, with 4a being created from 3 large datasets, and one small; 4b was created from three small datasets and 1 large; and 4c was created from two small and two large datasets. The times taken to populate each fact table were 173 ms, 210 ms and 188 ms respectively, with 4b (bik, lus, mot and ril) taking the largest time. This fact table was constructed from 2 JSON sources, 1 XML source and 1 HTML source while 4c (which was constructed from 2 JSON files, 1 CSV file and 1 XML file) proved to be faster by 22 ms. Once again, the time difference between 4b and 4c indicates that the time taken to populate a data mart is largely due to the format of the source data streams. We can see this if we compare the times taken for both 4b and 3b as they both used the same data sources (apart from the HTML source lus. With 3b being populated in 88 ms, and the addition of a fourth source lus as shown in 4b increases the population time by 122 ms.

Columns 5a and 5b show the times taken and the number of rows constructed for schemas created from 5 data sources, with 5a being created from 3 large datasets and 2 small ones and 5b being constructed from 3 small datasets and 2 large ones. Interestingly at this point, it appears that the number of instances created appears to surpass the initial bottlenecks provided by the source data formats, with 5a being populated in 310 ms and 5b in 254 ms. In short, as the number of data sources increases, the largest factor on population time, changes from the formats of the source schemas, to the number of rows produced by the population.

Finally, column 6 shows the time taken to populate the schema created from all 6 sources. The population time for all sources was 350 ms and resulted in a fact table of 1277 rows. Undoubtedly this configuration yielded the largest results for both population time and rows produced, when compared to 5a which is made from all data sources apart from ril the difference in time between the two is 40 ms, while the differing number of rows is only 2 rows. This indicates that, as the number of data sources increases, the number of rows new instances has a noticeable impact on performance.

With respect to the individual data sources presented. The inherent complexity of each schema does not seem to pose a problem during population, most likely due to the fact that the mappings file details the exact location of the data in each stream which is required for population. Overall, the time taken to populate a fact table from combined data sources is not so much limited by the number of individual facts (rows) per data source, but rather the format and structure of the source data provided. With formats such as XML and HTML taking a longer time to process than those which are published as JSON or CSV files. However, it appears that as the number of sources involved in the fact table population increases, the resultant number of instances to be created becomes the dominant factor in population time rather than the format of the data stream.

In summary, our system integrates and makes available in close to real-time the data produced by the different transport services, and experiments show a linear rise in times with an increase in the size of the update. The evaluation also demonstrated that:

- XML and HTML sources (bold font in Table 2 increase the cost of updates (157 ms for only 4 new instances in 2c)
- The HTML dataset (lus) performed notably worse than XML showing that more structured data performs better (5a *vs* 5b).
- Larger data marts with more data sources have less of an effect than an increase in new data instances (3b *vs* 2c).

Table 2. Update times for various data mart configurations

Config ID	2a	2b	2c	3a	3b	3c	4a	4b	4c	5a	5b	6
Souces	bus bik	bus **lus**	**lus ril**	bus bik rds	bus bik mot	**lus** mot **ril**	bus bik rds mot	bik **lus** mot **ril**	bus mot rds **ril**	bus bik **lus** mot rds	bus **lus** mot rds **ril**	bus bik **lus** mot rds **ril**
Population (ms)	91	76	157	156	88	159	173	210	188	310	254	350
Rows	346	270	4	1248	371	29	1273	130	1174	1275	1176	1277

6   Conclusions

In this paper, we demonstrated how StarGraphs can be used to create the ConstellationGraph which serves as a Data Mart in decision support systems. In addition, we provide a fast method for incorporating new data sources as they come online and our evaluation shows that these sources are in our data marts (almost) immediately after update. Our future work involves a project that uses 200 Agri (agricultural) data sources to provide more of a scalable stress test for this system. Secondly, we are investigating how the system can suggest to the user which combinations of datasets should be incorporated into the same data mart.

References

1. Baazizi, M.A., Lahmar, H.B., Colazzo, D., Ghelli, G., Sartiani, C.: Schema inference for massive JSON datasets. In: Extending Database Technology (EDBT) (2017)
2. Barnaghi, P.M., Bermúdez-Edo, M., Tönjes, R.: Challenges for quality of data in smart cities. J. Data Inform. Qual. **6**(2–3), 6:1–6:4 (2015)
3. Bellini, P., Benigni, M., Billero, R., Nesi, P., Rauch, N.: Km4city ontology building vs data harvesting and cleaning for smart-city services. J. Vis. Lang. Comput. **25**(6), 827–839 (2014)
4. Cappellari, P., Virgilio, R., Maccioni, A., Roantree, M.: A path-oriented RDF index for keyword search query processing. In: Hameurlain, A., Liddle, S.W., Schewe, K.-D., Zhou, X. (eds.) DEXA 2011. LNCS, vol. 6861, pp. 366–380. Springer, Heidelberg (2011). doi:10.1007/978-3-642-23091-2_31
5. Dublin Bikes API. (originally available in XML, now only available in JSON): http://api.citybik.es/v2/networks/dublinbikes
6. Dublin Bus RTPI. http://dublinked.ie/real-time-passenger-information-rtpi/
7. Dublin City travel times. http://opendata.dublincity.ie/TrafficOpenData/CP_TR/trips.csv
8. Gea, T., Paradells, J., Lamarca, M., Roldan, D.: Smart cities as an application of internet of things: experiences and lessons learnt in Barcelona. In: Seventh International Conference on Innovative Mobile and Internet Services in Ubiquitous Computing (IMIS), 2013, pp. 552–557. IEEE (2013)
9. Hernández-Muñoz, J.M., Vercher, J.B., Muñoz, L., Galache, J.A., Presser, M., Hernández Gómez, L.A., Pettersson, J.: Smart cities at the forefront of the future internet. In: Domingue, J., et al. (eds.) FIA 2011. LNCS, vol. 6656, pp. 447–462. Springer, Heidelberg (2011). doi:10.1007/978-3-642-20898-0_32
10. Irish Rail. http://api.irishrail.ie/realtime/index.htm?realtime_irishrail
11. Komamizu, T., Amagasa, T., Shaikh, S.A., Shiokawa, H., Kitagawa, H.: Towards real-time analysis of smart city data: a case study on city facility utilizations. In: IEEE 18th International Conference HPCC/SmartCity/DSS, pp. 1357–1364. IEEE (2016)
12. Nebot, V., Berlanga, R., Pérez, J.M., Aramburu, M.J., Pedersen, T.B.: Multidimensional integrated ontologies: a framework for designing semantic data warehouses. In: Spaccapietra, S., Zimányi, E., Song, I.-Y. (eds.) Journal on Data Semantics XIII. LNCS, vol. 5530, pp. 1–36. Springer, Heidelberg (2009). doi:10.1007/978-3-642-03098-7_1
13. Roantree, M., Liu, J.: A heuristic approach to selecting views for materialization. Softw. Pract. Exp. **44**(10), 1157–1179 (2014)
14. Roantree, M., McCann, D., Moyna, N.: Integrating sensor streams in phealth networks. In: 14th IEEE International Conference on Parallel and Distributed Systems, ICPADS 2008, pp. 320–327. IEEE (2008)

15. Scriney, M., O'Connor, M.F., Roantree, M.: Generating cubes from smart city web data. In: Proceedings of the Australasian Computer Science Week Multiconference, ACSW 2017, Australia, pp. 49:1–49:8 (2017)
16. Skoutas, D., Simitsis, A.: Ontology-based conceptual design of etl processes for both structured and semi-structured data. Int. J. Semant. Web Inform. Syst. (IJSWIS) **3**(4), 1–24 (2007)
17. Transport for Ireland. http://www.transportforireland.ie/real-time/real-time-ireland/
18. Transport Infrastructure Ireland. https://www.tiitraffic.ie/travel_times/

Towards Automatic Data Format Transformations: Data Wrangling at Scale

Alex Bogatu[✉], Norman W. Paton, and Alvaro A.A. Fernandes

School of Computer Science, University of Manchester, Manchester M13 9PL, UK
{alex.bogatu,npaton,a.fernandes}@manchester.ac.uk

Abstract. Data wrangling is the process whereby data is cleaned and integrated for analysis. Data wrangling, even with tool support, is typically a labour intensive process. One aspect of data wrangling involves carrying out format transformations on attribute values, for example so that names or phone numbers are represented consistently. Recent research has developed techniques for synthesising format transformation programs from examples of the source and target representations. This is valuable, but still requires a user to provide suitable examples, something that may be challenging in applications in which there are huge data sets or numerous data sources. In this paper we investigate the automatic discovery of examples that can be used to synthesise format transformation programs. In particular, we propose an approach to identifying candidate data examples and validating the transformations that are synthesised from them. The approach is evaluated empirically using data sets from open government data.

Keywords: Format transformations · Data wrangling · Program synthesis

1 Introduction

Data wrangling is the process of data collation and transformation that is required to produce a data set that is suitable for analysis. Although data wrangling may be considered to include a range of activities, from source selection, through data extraction, to data integration and cleaning [5], here the focus is on *format transformations*. Format transformations carry out changes to the representation of textual information, with a view to reducing inconsistencies.

As an example, consider a scenario in which information about issued building permits is aggregated from different data sources. There can be different conventions for most of the fields in a record (e.g. the format of the date when the permit was issued: *2013-05-03* vs *05/03/2013*; the cost of the building: *83319* vs *$83319.00*; or the address of the building: *730 Grand Ave* vs *Grand Ave, Nr. 730*). Such representational inconsistencies are rife within and across data sources, and can usefully be reduced during data wrangling.

Data wrangling is typically carried out manually (e.g. by data scientists) with tool support; indeed data wrangling is often cited as taking a significant

© Springer International Publishing AG 2017
A. Calì et al. (Eds.): BICOD 2017, LNCS 10365, pp. 36–48, 2017.
DOI: 10.1007/978-3-319-60795-5_4

portion of the time of data scientists[1]. Data scientists can author format transformations manually, but tool support is available. For example, in Wrangler [8], commercialised by Trifacta, data scientists author transformation rules with support from an interactive tool that can both suggest, and illustrate the effect of, the rules. Such an approach to data wrangling should lead to good quality results, but is labour intensive where there are multiple data sources that manifest numerous inconsistencies.

In this paper we address the question *can the production of such format transformations be automated?* Automatic solutions are unlikely to be able to match the reach or quality of transformations produced by data scientists, but any level of automation provides the possibility of added value for minimal cost. We build on some recent work on the synthesis of transformation programs from examples, which was originally developed for use in spreadsheets (e.g. Flash-Fill [6], BlinkFill [12]). In the commercial realisation of FlashFill as a plugin to Excel, the user provides example pairs of values that represent source and target representations, from which a program is synthesised that can carry out the transformations. The published evaluations have shown that effective transformations can often be produced from small numbers of examples.

An issue with this is that there is a need for examples to be provided by users. While this is potentially fine for spreadsheets, where there is typically a single source and target, and the source is of manageable size, identifying suitable examples seems more problematic if there are large data sets or many sources. How do we scale program synthesis from examples to work with numerous sources? The approach investigated here is to identify examples automatically. Specifically, the main contributions of this paper are: (i) the identification of the opportunity to deploy format transformation by program synthesis more widely through automatic discover of examples; (ii) the description of an approach that supports (i); and (iii) the evaluation of the resulting approach with real world data sets. Although this paper focuses on a fully automated approach, in which the user is not in-the-loop, the automated approach could be used to support the bootstrapping phase of pay-as-you-go approaches, in which users subsequently provide feedback on the results of the automatically synthesised transformations.

2   Technical Context

In this section, we briefly review the work on synthesis of programs for data transformation on which we build; full details are provided with the individual proposals (e.g. [6,12]). In essence, the approach has the following elements: (i) a *domain-specific* language within which transformations are expressed; (ii) a *data structure* that can be used to represent a collection of candidate transformation programs succinctly; (iii) a program that can generate candidate transformation programs that correspond to examples; and (iv) a ranking scheme that supports the selection of the more general programs. This language can express a range

---

[1] http://nyti.ms/1Aqif2X.

Table 1. Transformation scenario

	Address	Transformed address
1	730 Grand ave	Grand Ave, Nr. 730
2	5257 W Eddy St	W Eddy St, Nr. 5257
3	362 Schmale Rd	
4	612 Academy Drive	
5	3401 S Halsted Rd	

of operations for performing syntactic manipulations of strings such as *concatenation* - **Concatenate** which links together the results of two or more *atomic expressions* - **SubStr, Pos**, or more complex expressions that involve *conditionals* - **Switch** and *loops* - **Loop**. The user provides the synthesizer a number of examples, which is usually small, depending on the complexity of the values to be transformed. Then, the synthesizer generates a set of programs expressed using the functions above and consistent with the values provided as examples. The most suitable programs are chosen based on the ranking scheme.

To illustrate the approach in practice, in Table 1, the user provided the first two rows as examples for transforming the Address column. Given the examples, the synthesizer will try to learn one or more programs consistent with the values provided by the user. For instance, for row 1, in FlashFill, the following expression is a possible inferred program:

Concatenate$(e_1,$ **ConstStr**$(", Nr."), e_2)$, where
$e_1 \equiv$ **SubStr**$(v_1,$ **Pos**$($AlphTok, ϵ, 1$),$ **Pos**$($EndTok, ϵ, 1$)),$
$e_2 \equiv$ **SubStr2**$(v_1,$ NumTok, 1$),$ and
$v_1 \equiv$ the left-hand side value (e.g. "730 Grand Ave")

The logic here is to extract the street name as e_1 using the **SubStr** function, i.e., the sub-string that starts at the index specified by the first occurrence of an alphabet token and ends at the end of string, then to extract the street number as e_2 using the **SubStr2** function, i.e., the first occurring number, and concatenate these two sub-strings separated by the constant string ", Nr. " using the **Concatenate** function. This program is consistent with the second example as well, and when applied on the rest of the values (rows 3–5) it will transform them into the desired format.

While the above example illustrates a case in which the program learned is able to correctly transform all the values, this is not always the case. In general, the generated transformations are more likely to succeed to the extent that the following hold: (a) the correct transformation is expressible in the underlying transformation language, (b) both elements in the example pairs denote values in the domain of the same real-world property, and (c) taken together, the example pairs cover all or most of the formats used for the column.

S.Permit_Nr.	S.Date	S.Contractor	S.Cost	S.Address	S.Permit_Type
100484472	2013-05-03	BILLY LAWLESS	83319	730 Grand Ave	Renovation

(a) Source

T.Permit_Nr.	T.Date	T.Cost	T.Contractor	T.Address	T.Type
100484472	05/03/2013	$83319.00	BILLY LAWLESS	Grand Ave, Nr. 730	Renovation

(b) Target

$Matches(S,T)$	$\{\langle S.Permit\_Nr., T.Permit\_Nr.\rangle, \langle S.Date, T.Date\rangle,$ $\langle S.Cost, T.Cost\rangle, \langle S.Address, T.Address\rangle, ...\}$
$S.FD = T.FD$	$\{Permit\_Nr. \rightarrow Date, Permit\_Nr. \rightarrow Cost,$ $Permit\_Nr. \rightarrow Address, Permit\_Nr. \rightarrow Contractor, ...\}$
Generated Examples	$\{\langle S.Date, T.Date, \langle"2013-05-03", "05/03/2013"\rangle\rangle,$ $\langle S.Cost, T.Cost, \langle"83319", "\$83319.00"\rangle\rangle,$ $\langle S.Address, T.Address, \langle"730\ Grand\ Ave", "Grand\ Ave,\ Nr.\ 730"\rangle ...\}$

(c) Partial intermediate and final results from the algorithm

Fig. 1. Building permits example

3   Discovering Examples

The approach described in Sect. 2 synthesises transformation programs from examples, where an *example* consists of pairs of source and target values, $\langle s, t \rangle$, where s is a literal from the source and t is a literal from the target. In our running example, $s = 730\ Grand\ Ave$ and $t = Grand\ Ave,\ Nr.\ 730$. In the spreadsheet setting, given a column of source values, the user provides target examples in adjacent columns, and FlashFill synthesises a program to transform the remaining values in ways that are consistent with the transformations in the examples. Of course, it may take several examples to enable a suitable transformation (or suitable transformations) to be synthesised. In fact, the user needs to provide enough examples to cover all the relevant patterns existing among the values to be transformed.

In this section we propose an approach to the automatic identification of examples, drawing on data from existing data sets. Our aim is to use techniques such as the one described in the previous section, in more complex scenarios than spreadsheets. For example, consider a scenario in which we would like to integrate information about issued building permits from several different sources, and for the result to be normalized. Specifically, we want to represent the columns of the resulting dataset using the formatting conventions used in one of the original datasets, which acts as the target. Providing manual examples to synthesize the transformations needed can be a tedious task that requires knowledge about the formats of the values existing in the entire dataset. In our approach, the basic idea is to identify examples from the different sources, where the source and target values for an attribute can be expected to represent the same information.

To continue with our running example, assume we have two descriptions of an issued building permit, as depicted in Fig. 1(a) and (b). To generate a transformation that applies to the *Address* columns, we need to know which (different) values in the source and target *Address* columns are likely to be equivalent. There are different types of evidence that could be used to reach such a conclusion. In the approach described here, we would draw the conclusion that *730 Grand Ave* and *Grand Ave, Nr. 730* are equivalent from the following observations: (i) the names of the first columns in the two tables match (because of the identical sub-string *Permit_Nr.* they share); (ii) there is a functional dependency, on the instances given, *Permit_Nr.* → *Address* in each of the tables; (iii) the names of the fifth columns (*Address*) of the two tables match; and (iv) the values for the first columns in the two tuples are the same. Note that this is a heuristic that does not guarantee a correct outcome – it is possible for the given conditions to hold, and for the values not to be equivalent; for example, such a case could occur if the *Address* attributes of the source and target tables had different semantics.

More formally, assume we have two data sets, source S and target T. S has the attributes $(sa_1, ...sa_n)$, and T has the attributes $(ta_1, ...ta_m)$. We want values from S to be formatted as in T. Further, assume that we know instances for S and T. Then we can run a functional dependency discovery algorithm (e.g. [9]) to hypothesize the functional dependencies that exist among the attributes of S and T. This gives rise to collections of candidates functional dependencies for S and T, $S.FD = \{sa_i \rightarrow sa_j, ...\}$ and $T.FD = \{ta_u \rightarrow ta_v, ...\}$. Note that, in general, sa_i and ta_u can be lists of attributes.

In addition, assume we have a function, *Matches*, that given S and T, returns a set of pairwise matches between the attribute names in S and T, $Matches(S,T) = \{\langle sa_i, ta_j \rangle, ...\}$. *Matches* can be implemented using a schema matching algorithm, most likely in our context making use of both schema and instance level matchers [10]. In the case where the left hand sides of the two functional dependencies $sa_i \rightarrow sa_j$ and $ta_u \rightarrow ta_v$ are lists of attributes, we say that sa_i *matches* ta_u if both lists have the same number of elements and there are pairwise matches between the attributes of the two lists. Then Algorithm 1 can be used to compute a set of examples for transformations between S and T. For the *join condition* used in the SQL query at line 7, if $S.sa$ and $T.ta$ are list of attributes, then the equality is true if both lists have the same number of elements and there are pairwise equalities between the matching attributes of the two lists. While this may seem restrictive, it is a necessary condition for finding suitable example candidates.[2]

---

[2] It can be seen that the complexity of Algorithm 1 is $\mathcal{O}(nm)$ where n is the number of attributes of S and m is the number of attributes of T. This is due to the cross product between the columns of the two data sets (i.e. the two for loops from the beginning of the algorithm). We do not analyse here the complexity of the other algorithms used in our experiments as this has been done in the original papers. Nor do we emphasize on the impact of input size on the overall solution. In our experiments, the run-time of Algorithm 1, pertaining examples generation alone, did not exceed one second for any of the datasets used.

Algorithm 1. Example discovery using functional dependencies.

1: **function** FINDEXAMPLES(S,T)
2:     $Examples \leftarrow \{\}$
3:     **for all** $sa \in S$ **do**
4:         **for all** $ta \in T$ **do**
5:             **if** $\langle sa, ta \rangle \in Matches(S, T)$ **then**
6:                 **if** $(sa \rightarrow sad) \in S.FD$ and $(ta \rightarrow tad) \in T.FD)$ and $\langle sad, tad \rangle \in$
        $Matches(S, T)$ **then**
7:                     $EgVals \leftarrow$ select distinct S.sad, T.tad from S, T where
        S.sa = T.ta
8:                         $Examples \leftarrow Examples \cup \langle sad, tad, EgVals \rangle$
9:                     **end if**
10:                **end if**
11:            **end for**
12:        **end for**
13:        **return** $Examples$
14: **end function**

For the example in Fig. 1(a) and (b), (c) illustrates the intermediate results used by or created in Algorithm 1.

Although Algorithm 1 returns sets of examples that can be used for synthesising transformations (for example, using FlashFill [6] or BlinkFill [12]) there is no guarantee that the transformation generation algorithm will produce effective transformations. The synthesised transformations may be unsuitable for various reasons, e.g., (a) the required transformation cannot be expressed using the available transformation language, (b) the data is not amenable to homogenisation (e.g. because there is no regular structure in the data), or (c) there are errors in the data. As a result, there is a need for an additional validation step that seeks to determine (again automatically) whether or not a suitable transformation can be synthesised.

In our approach, the set of examples returned by Algorithm 1 is discarded unless a k-fold cross validation process is successful. In this process, the set of examples is randomly partitioned into k equally sized subsets. Then, transformations are synthesised, in k rounds, using the examples from the other $k - 1$ partitions, and the synthesised transformation is tested on the remaining partition. Although k-fold cross validation can be used with different thresholds on the fraction of correct results produced, in the experiments, we retain a set of examples only if the synthesised transformations behave correctly throughout the k-fold validation process. In our experiments, we used $k = 10$.

4   Evaluation

In this paper we hypothesise that the process of transforming data from one format into another using the recent work on program synthesis can be automated by replacing the user-provided examples with those identified by Algorithm 1. In a scenario in which information from multiple, heterogeneous data sets is to be integrated, in Wrangler [8] or FlashFill [6], it falls on the data scientists

Table 2. Data sources used in experiments

Domain	URLs	Cardinality	Arity
Food Hygiene	http://ratings.food.gov.uk/,	12323	7
	https://data.gov.uk	61	10
Lobbyists	https://data.cityofchicago.org/,	7542	12
	https://data.cityofchicago.org/ (different years)	210	19
Doctors/addresses	http://www.nhs.uk	7696	19
	https://data.gov.uk	17867	3
Building permits	https://app.enigma.io,	10000	13
	https://data.cityofchicago.org/	1245	13
Citations	http://dl.acm.org/results.cfm,	1072	16
	http://ieeexplore.ieee.org/	2000	11

either to identify values that need to be transformed in one data set and their corresponding versions in a second data set, or to provide the target versions for some of the source values. Our approach removes the user from the process by automating the identification of examples for use by program synthesis. To illustrate and evaluate this method, we build a process pipeline which makes use of several existing pieces of work. Specifically, the process starts by using an off-the-shelf matcher to identify column level matches between a source data set and a target one (*Match*), then for each data set an off-the-shelf profiling tool is used to identify functional dependencies, which are then used together with the previously discovered matching column pairs to generate input-output examples for the desired transformations using Algorithm 1. The sets of examples that result are validated using a k-fold cross-validation process as described in Sect. 3. Finally, the validated pairs are fed to a synthesis algorithm. Note that only the generation of examples is claimed as our main contribution and, hence, evaluated in this section. Although they are relevant to the use of the approach in practice, we do not seek to directly evaluate the off-the-shelf components we use: (i) the effectiveness of the implementation of *Match*; (ii) the effectiveness of the functional dependency detector; or (iii) the effectiveness of the format transformation synthesizer. In all cases, these works have been evaluated directly by their original authors and/or in comparative studies, and we do not seek to replicate such studies here. Rather, we carry out experiments that investigate example discovery, validation and the quality of the synthesised program, for the approach described in Sect. 3.

Experimental Setup. In the experiments we use data from the open government data sets listed in Table 2. For each domain, the URLs represent the location of the source and target datasets, respectively. The last two columns illustrate the size (cardinality and arity) of each dataset. For Matching, we used

COMA 3.0 Community Edition[3], with schema and instance level matchers, and the following parameter values: $MaxN = 0$ and $Threshold = 0.2$. The first parameter specifies that attributes can only be in 1:1 matching relationships with each other. The second value specifies the minimal confidence a correspondence between two elements must reach in order to be accepted as a match. For functional dependency discovery we used HyFD[4] [9] with the *Null Semantics* setting set to $null \neq null$. This last setting is needed because real-world data often contains *null* values. So for a schema $S(A, B)$, if there are two tuples $r1 = (null, 1)$ and $r2 = (null, 2)$, if $null = null$ then $A \rightarrow B$ is not a functional dependency. In order to avoid discarding functional dependencies in such situations we assume $null \neq null$. For synthesising and running transformations, we used the Flash-Fill feature from Excel 2013. All experiments were run on Ubuntu 16.04.1 LTS installed on a 3.40 GHz Intel Core i7-6700 CPU and 16 GB RAM machine.

Experiment 1. *Can Algorithm 1 identify candidate column pairs?* To evaluate this, we report the precision ($\frac{TP}{TP+FP}$) and recall ($\frac{TP}{TP+FN}$) of the algorithm over data sets where there are known suitable candidates. For this experiment, *true positive* (TP), *false positive* (FP) and *false negative* (FN) are defined as: TP – the two columns represent the same concept; FP – the two columns do not represent the same concept; and FN – a correct result that is not returned. Examples of candidate columns pairs that are identified by Algorithm 1 as examples, are given in Table 3.

The results for this experiment are presented in Table 4, for the data sets in Table 2. In general, both precision and recall are high. In the case of the Building Permits domain, there are 3 attributes in each dataset representing a cost (see Row 2 in Table 3 for an example). Although there are 9 possible source column-target column alignments, Algorithm 1 was able to identify the correct ones and returned no false positives. Food Hygiene precision is *0.86* due to a FP match identified by COMA. An example of this match is on the second row of Table 3. The two columns have the same name (*AddressLine1*), but different semantics. The precision and recall for Citation are reduced by one FP match identified by COMA (represented in the last row of Table 2) and 2 FNs (i.e. 2 pairs of columns that were not reported as matches by COMA).

Experiment 2. *Is the validation process successful at identifying problematic column pairs?* A column pair is problematic if we cannot synthesise a suitable transformation. To evaluate this, we investigate the candidate pairs for which validation has failed; the cases for which validation has passed are discussed in Experiment 3. As described in Sect. 3, the validation step consists of a k-fold cross validation with $k = 10$. A set of examples is considered to pass this validation step if all transformations are correct for each of the 10 iterations.

The fraction of the candidate column pairs that pass validation is reported in Table 4. Column pairs have failed validation for the following reasons: (i) The matched columns have different semantics, and thus incompatible values,

---

[3] http://bit.ly/2fLVvtl.
[4] http://bit.ly/2f5DwJW.

Table 3. Transformation candidates

#	Domain	Src semantics	Source example	Target semantics	Tgt example	# egs.
1	Food Hyg.	Rating date	2015-12-01	Rating date	01/12/2015	57
2	Food Hyg.	Building name	Royal free hospital	Street	Pond street	59
3	Food Hyg.	Business name	Waitrose	Business name	Waitrose	60
4	Lobbyists	Person name	Mr. Neil G Bluhm	Person name	Bluhm Neil G	206
5	Lobbyists	Phone Nr.	(312) 463–1000	Phone Nr.	(312) 463–1000	191
6	Docs./Addrs.	Address	55 Swain Street, Watchet	Street	Swain Street	41
7	Docs./Addrs.	City	Salford	City	Manchester	28
8	Build. Permits	Address	1885 Maud ave	Address	Maud Ave, Nr. 1885	26
9	Build. Permits	Cost	6048	Cost	$6048.00	22
10	Build. Permits	Issue date	2014-06-05	Issue date	06/05/2014	26
11	Citations	Author names	Sven Apel and Dirk Beyer	Author names	S. Apel; D. Beyer	56
12	Citations	Conf. date	””	Conf. date	2–8 May 2010	32
13	Citations	Pub. year	2011	Pub. year	2011	56
14	Citations	Nr. of pages	10	Start page	401	56

for which no transformation can be produced. This is the case for *2* of the *8*
column pairs that fail validation (for example see Row 2 and Row 14 in Table 3).
(ii) The matched columns have the same semantics, but FlashFill has failed to
synthesise a suitable transformation. This is the case for *2* of the column pairs
that fail validation (for example, consider the lists of author names from Row 11
in Table 3). (iii) There are issues with the specific values in candidate columns.
This is the case for *4* of the column pairs that fail validation (as an example
consider the missing information from Row 12 in Table 3) and inconsistent values
(in Row 7 in Table 3). It is important to note that in practice, the effectiveness
of the off-the-shelf tools that we used here can be impacted by characteristics
of the data such as the ones exemplified above. This evaluation shows that the
validation method we employ is able do filter out example sets that otherwise
would produce invalid transformations or no transformations at all.

Experiment 3. *Do the synthesised transformations work on the complete data
sets (and not just the training data)?* To evaluate this, we report the precision
and recall of validated transformations in Table 5; the missing row numbers are

Table 4. Experiments 1 and 2

Domain	# of candidates	Alg. precision	Alg. recall	Fraction validated
Food Hyg.	6	0.83	1.00	5/6
Lobbyists	9	1.00	1.00	9/9
Docs./Addrs.	2	1.00	1.00	1/2
Build. Perm.	12	1.00	1.00	12/12
Citations	7	0.86	0.75	1/7

the examples from Table 3 for which the transformations failed validation. In computing the precision and recall, we use the following definitions: TP – the transformation produces the correct output; FP – the transformation produces an incorrect output; FN – the transformation results in an empty string.

Of the *28* validated transformations from Table 4, all but *6* are identity transformations, i.e. the source and target data values are the same (e.g. Rows 3, 5 and 13 in Table 5). This can often happen in practice. For instance, Row 13 represents the year for a publication which is most commonly represented as a four digit number. In such cases, FlashFill proved to be able to identify that the transformation is only a copying operation. Of the *6* cases where the values are modified, the precision and recall are both *1.0* in 3 cases (Rows 1, 8 and 10 in Table 5). For rows 1 and 10 the transformations rearrange the date components and replace the separator. Our experiments confirmed the results of [6] according to which FlashFill is effective in such simple cases. For Row 8, the transformation is more complicated given that the street name does not always have a fixed number of words, or that the street number can have several digits. In this case Algorithm 1 was able to identify enough examples to cover all the existing formats. There were problems with the transformations generated in 3 cases. For Row 4, a few source values do not conform to the usual pattern (e.g. the full stop is missing after the salutation). For Row 9, not all source values are represented as integers, giving rise to incorrect transformations. For Row 6, similarly to Row 9, the examples do not cover all the relevant address formats, i.e. 41 examples are used to synthesize a program to transform a rather large number of values (approx. 7700).

Conclusion. The technique evaluated above proves to be effective in scenarios where certain conditions are met. The most important of these is that the source and target contain overlapping information on some of the tuples, i.e. the left-hand sides of the functional dependencies used in Algorithm 1, and format diversity on the corresponding right-hand side. Another important condition is for FlashFill to be able to synthesise transformations from the pairs of generated examples. The evaluation shows that as long as these conditions are met, we can delay the need for user intervention in the cleaning process by synthesizing and applying some transformations automatically. Although the synthesis algorithms suffer from some limitations as mentioned in Sect. 2, we leave the opportunity of addressing these shortcomings for future work and use FlashFill as a black-box in this paper.

Table 5. Experiment 3

#	Src semantics	Src value	Tgt semantics	Tgt value	Precision	Recall
1	Rating date	2015-12-01	Rating date	01/12/2015	1.0	1.0
3	Business name	Waitrose	Business name	Waitrose	1.0	1.0
4	Person name	Mr. Neil G Bluhm	Person name	Bluhm Neil G	0.98	0.84
5	Phone Nr.	(312) 463–1000	Phone Nr.	(312) 463–1000	1.0	1.0
6	Address	55 Swain Street, Watchet	Street	Swain street	0.68	1.0
8	Address	1885 maud ave	Address	Maud ave, Nr. 1885	1.0	1.0
9	Cost	6048	Cost	$6048.00	0.97	1.0
10	Issue date	2014-06-05	Issue date	06/05/2014	1.0	1.0
13	Pub. year	2011	Pub. year	2011	1.0	1.0

5   Related Work

In recent years, there has been an increasing number of proposals that use declarative, constraint-based quality rules to detect and repair data problems (e.g. [1,4,13,15], see [2,3] for surveys). For many of these heuristic techniques rule-based corrections are possible as long as the repairing values are present either in an external reference data set or in the original one. For example, in [4] the correct values are searched for in a master data set using editing rules which specify how to resolve violations, with the expense of requiring high user involvement. While the semantics of editing rules can be seen as very similar to our approach described in Algorithm 1, there are at least two essential differences. First, editing rules address instance-level repairs, i.e. every tuple is checked against every rule and the values of the attributes covered by the rule are replaced with correct ones from the reference data set (if they exist). Our approach determines pairs of values from which we learn transformations that hold for entire columns, so we do not search for the correct version of every value that needs to be cleaned, but for a small number of values that describe the syntactic pattern of the correct version. Second, we determine these transformations automatically, without any intervention from the user.

Closer to our solution are the tools for pattern enforcement and transformation, starting with traditional ETL tools like Pentaho[5] or Talend[6], but especially Data Wrangler [8], its ancestor Potter's Wheel [11], and OpenRefine[7].

---

[5] http://www.pentaho.com/.
[6] https://www.talend.com/.
[7] http://openrefine.org/.

Data Wrangler stands out by proposing a transformation language and an inference algorithm that aid the user in transforming the data. Although the automated suggestion mechanism, described in [7], avoids manual repetitive tasks by actively learning from the decisions the user makes, or from the transformations the user writes, the user must know what transformation is required and how to express that operation. Our work is a first step towards a solution in which such transformations are synthesised without up-front human involvement.

An important body of related work is the research on synthesis programming introduced in Sect. 2. The algorithms and languages proposed in [6,12], and extended in [14] have been developed with spreadsheets scenarios in mind. We build on these solutions and argue that such techniques can be applied on real world big data as well, where the amount of inconsistency and format heterogeneity is higher.

6 Conclusions

Data wrangling is important, as a precursor to data analysis, but is often labour intensive. The creation of data format transformations is an important part of data wrangling and in the data cleaning landscape there are important recent results on techniques to support the creation of such transformations (e.g. [6,8]). These solutions are user-centric and many of the transformations that can be automatically learned from examples provided by Algorithm 1 can be obtained using the above solutions as well, but with a much greater level of human involvement. In this paper we build on and complement these results by describing an approach that can automate the creation of format transformations, where certain conditions are satisfied. Thus we do not replace human curators, but we reduce the number of cases in which manual input is required. In several domains, we have described how candidate sets of examples can be discovered automatically, how these examples can be used to synthesise transformations using FlashFill, and how the resulting transformations can be validated automatically. In experiments, the validated transformations have high precision.

Acknowledgement. This work has been made possible by funding from the UK Engineering and Physical Sciences Research council, whose support we are pleased to acknowledge.

References

1. Chu, X., Ilyas, I.F., Papotti, P.: Holistic data cleaning: putting violations into context. In: ICDE 2013, pp. 458–469 (2013)
2. Fan, W.: Dependencies revisited for improving data quality. In: PODS 2008, pp. 159–170, 9–11 June 2008
3. Fan, W.: Data quality: from theory to practice. SIGMOD Rec. **44**(3), 7–18 (2015)
4. Fan, W., Li, J., Ma, S., Tang, N., Yu, W.: Towards certain fixes with editing rules and master data. VLDB J. **21**(2), 213–238 (2012)

5. Furche, T., Gottlob, G., Libkin, L., Orsi, G., Paton, N.W.: Data wrangling for big data: challenges and opportunities. In: EDBT, pp. 473–478 (2016)
6. Gulwani, S.: Automating string processing in spreadsheets using input-output examples. In: POPL, pp. 317–330 (2011)
7. Heer, J., Hellerstein, J.M., Kandel, S.: Predictive interaction for data transformation. In: CIDR 2015, 4–7 January 2015
8. Kandel, S., Paepcke, A., Hellerstein, J., Heer, J.: Wrangler: interactive visual specification of data transformation scripts. In: CHI, pp. 3363–3372 (2011)
9. Papenbrock, T., Naumann, F.: A hybrid approach to functional dependency discovery. In: SIGMOD Conference, pp. 821–833. ACM (2016)
10. Rahm, E., Bernstein, P.A.: A survey of approaches to automatic schema matching. VLDBJ 10(4), 334–350 (2001)
11. Raman, V., Hellerstein, J.M.: Potter's wheel: an interactive data cleaning system. In: VLDB 2001, pp. 381–390, 11–14 September 2001
12. Singh, R.: BlinkFill: semi-supervised programming by example for syntactic string transformations. PVLDB 9(10), 816–827 (2016)
13. Jia, X., Fan, W., Geerts, F., Kementsietsidis, A.: Conditional functional dependencies for capturing data inconsistencies. TODS 33(1), 6:1–6:48 (2008)
14. Wu, B., Knoblock, C.A.: An iterative approach to synthesize data transformation programs. In: IJCAI 2015, pp. 1726–1732, 25–31 July 2015
15. Yakout, M., Elmagarmid, A.K., Neville, J., Ouzzani, M., Ilyas, I.F.: Guided data repair. PVLDB 4(5), 279–289 (2011)

Database Processes for Application Integration

Daniel Ritter[(✉)]

Faculty of Computer Science, University of Vienna, Vienna, Austria
`danielr81@univie.ac.at`

Abstract. Recent technology trends such as Cloud and Mobile Computing generate huge amounts of heterogeneous data to be processed by business and analytic applications. Currently, the data is stored in (relational) databases, before being queried by these applications. The heterogeneity of sender and application formats requires integration processes for decoupling sender from receiver (i. e., no shared schema), routing (i. e., "which application receives which parts of the data?") and format transformations. We argue that this kind of database (integration) process does not exist and propose a suitable processing model that fits the general integration semantics. The analysis is conducted for a real-world integration scenario on a well-established relational database system.

Keywords: Application integration · Database processes · Integration pattern semantics

1 Introduction

The advent of data generating trends (e. g., cloud computing, internet of things) on the one hand and the interest of business and analytic applications to process the data leads to several challenges (e. g., decoupled communication participants, routing, transformation) that are usually solved by enterprise application integration (EAI) systems [11,13]. In EAI, an ordered set of message processors are applied in the form of an integration process. At the same time, these integration systems — having to process huge amounts of data — are looking for faster processing techniques closer to the actual data.

We argue that not only could applications having their data in the same database system benefit from integration processes within the database system, but also that EAI systems could "push-down" their semantics — denoted by the enterprise integration patterns (EIPs) [11] — to the database in order to improve their processing [17]. There are three types of integration scenarios that would benefit from this [15]: *(C1)* two applications with data in the same database system with built-in integration logic (↦ internal/internal), *(C2)* one application with data on the database system (with integration logic) receives or sends out data (↦ internal/external), *(C3)* two applications that have no data on the database system with integration logic (↦ external/external). In previous work [16], we investigated *(C3)* on the application tier by "moving up" database

© Springer International Publishing AG 2017
A. Calì et al. (Eds.): BICOD 2017, LNCS 10365, pp. 49–61, 2017.
DOI: 10.1007/978-3-319-60795-5_5

processing techniques into an integration system. The results were especially promising for message throughput and scalability.

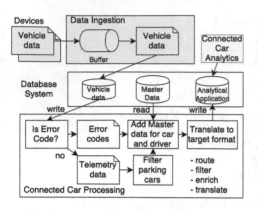

Fig. 1. CCT process with focus on the integration processing (white boxes).

In this work we address the internal/internal *(C1)* case and investigate the EIP semantics in the context of relational database processing. We explain and experimentally evaluate our approach using a connected cars telematics (CCT) scenario[1], e. g., used in vehicle logistics and insurance domains, which represents a new class of integration scenarios that are currently difficult to support in EAI systems. Figure 1 denotes a simplified version of CCT processing (cf. white boxes; case *(C1)*). Briefly, the `Vehicle Data` messages represent telemetry data (e. g., vehicle speed) and error codes that are received at high frequency and volume by a message `Buffer` (i. e., *(C2)*) and stored during the data ingestion into the vehicle data table (cf. case *(C2)*). We focus on the CCT processing that includes integration semantics like content-based routing (i. e., to distinguish between the telemetry and error code data), content filtering (i. e., removing unnecessary data), content enrichment (i. e., adding master data), and translation into the format of the target application. Furthermore, the message senders (`Devices`) are decoupled from the receivers (`Connected Car Analytics`). Besides the de-coupling of a growing number of distinct CCT senders, they produce varying workloads that are currently difficult to process by applications. Current solutions show a tight coupling between senders and the CCT application and are implemented in a process-like but "ad-hoc" manner and do not formalize or optimize the processing according to integration semantics. While *(C2)* is partially covered by JMS-like message queuing extensions of database systems [7] (i. e., `Buffer` to Vehicle data), integration processes in the database (cf. *(C1)*) were partially addressed by evaluating nested views for engineering applications in [9], and thus proposed by [2] as future research. We answer the following research questions:

- **Q1:** "How can EAI semantics be represented in database systems?" This includes basic EIP semantics like the representation of messages and message channels, as well as message processors that require extended semantics [16].
- **Q2:** "How to compatibly adapt EAI processing semantics to databases?" A transactional process model is required that is compliant with the standard integration processing.
- **Q3:** "Can databases accelerate message processing?" We measure message throughput using the EIPBench benchmark [18] for selected EIPs.

---

[1] e. g., SAP "Vehicles Insights", visited 04/2017: https://www.sap.com/product/analytics/vehicle-insights.html.

Our contributions are a formalization and adaptation of the EIP semantics for database processing (Sect. 2; answering *Q1*). This includes the representation of basic integration principles (e. g., document message, datatype channel, canonical data model) in Sect. 2.1 as well as two important EIPs (according to [18]), for which we define our semantics compared to the current EIPs [11] in Sect. 2.2. In Sect. 3 we specify the transactional processing semantics (answering *Q2*) by introducing the transactional process model in Sects. 3.1 and 3.3 and its compilation to database processes with transaction detection in Sect. 3.4. We evaluate our approach in Sect. 4 (answering *Q3*).

2   Relational Pattern Formalizations

In this section we introduce basic integration semantics and discuss the implications on the EIPs (all taken from [11], if not marked otherwise) and implementations for some of the required EIPs from the motivating example (e. g., message, message channel, message filter, content-based router, message translator).

2.1   Basic Principles: Multi-relational Message Collections

Set-oriented processing changes the structure of messages, however, does not require any change at the message channel level — although the current document message or the datatype channel could be specialized to indicate the nature of the relational messages exchanged. A message $m:=(b, H_b)$ according to [11] consists of a body b with arbitrary content and a set of name/value-pair meta-data entries describing b, called the header H_b. An extension for additional multimedia data — not in [11] — is a set of name/binary value attachments A.

For example, an e-Mail Adapter is able to receive messages with a textual body and attachments. For relational message processing, the body b has to be restricted to relational message contents. For example, Fig. 2 denotes the body b_r, header H_{r,b_r} and attachment A_r relations. In addition, a control record for each message is inserted into an optional message relation, which manages the mapping of messages to faults that happened during processing. As such, m_r denotes a

Message (optional)		Body			
Message ID <int>	fault <int>	Message ID <int>	Data Col 1 <type 1>	...	Data Col N <type N>
1	0	1	'#1'	...	'invoice'
2	0	2	'#1'	...	'order'

Header (optional)			Attachments (optional)		
Message ID <int>	Key <string>	Value <string>	Message ID <int>	Key <string>	Value <blob>
1	sender	IBM	1	invoice	<doc>
2	sender	CISCO	2	order	<doc>

Fig. 2. Multi-relational message collection

"multi-relational" message, since it has to be represented using one relation for each of the distinct formats of the message entries. A relational message $m_r :=$ (b_r, H_{r,b_r}, A_r) consists of a relational message body b_r in normalized form, and sets of headers H_{r,b_r} and attachments A_r represented by ternary relations in the form of name/value character, and name/binary character-clob/blob fields. When considering b_r to be in BCNF, a relational message m_r consists of at least one and up to n relations, allowing for different, possibly joining body relations. In conventional message processing, the body table would have only one entry in each of the corresponding database tables, representing the content of one message. For more efficient processing in databases, we define a relational message collection MC_r of m_r. A (relational) message collection $MC_r :=$ $(B_r^{f_1}, B_r^{f_2}, ... B_r^{f_k}, H_{r,b_r}, A_r)$ consists of a collection of k message content sets B^f of the same message format f, H_{r,b_r} and A_r relations. While the header and attachment relations can be shared between messages, the body might require additional relations, if the message formats differ.

In practice, the constructed MC_r can be represented by a table instance for each relation. With this representation tree- and graph like message formats (e. g., XML, JSON) can be expressed. However, the header and attachment relations, which are added as input and output to each message processor, are accessible using relational algebra (e. g., selection, join) in contrast to standard XML processing (e. g., not considered in [4]).

2.2   Integration Patterns for Relational Processing by Example

The pipeline model of integration processes consists of message channels that connect message processors by passing the outgoing message of the previous processor to the subsequent processor as input message [11]. These processors are generally categorized as routing and transformation patterns. Due to brevity, we focus on two of the patterns from the motivating example (cf. Fig. 1), namely a content-based router (incl. message filter) and a message translator. For a general expressiveness discussion of relational EIPs, see [15,16].

Content-based Router: The semantics of the content-based router [11] are denoted in Listing 1.1. Note that a router with only one outgoing channel and one condition is called message filter [11]. With a message cardinality of 1:1 and channel cardinality of 1:n (actually 1:1, since only one channel is selected) the router checks the message content sequentially (cf. line 2). If a channel condition $ch_i.cond()$ matches the original message m, then m is routed to that channel and no further condition is examined (cf. line 3). Otherwise, the other channels are checked in sequence until one matches. The message is routed to a default channel $ch_{default}$, if none of the conditions match (cf. line 6). Since the original message is forwarded, the router is read-only and non-message generating.

Listing 1.1. Current CBR semantics

```
sequential for channel ch_i ∈ C, m
    if match(ch_i.cond(), m) then
        route(m, ch_i); return;
    end if;
end for;
route(m, ch_default);
```

Listing 1.2. Table CBR semantics

```
parallel for channel ch_i ∈ C;
    create(MC_r^{ch_i})
    for message m_{r,j} ∈ MC_r
        if match(ch_i.cond(), m_{r,j}) then
            add(m_{r,j}, MC_r^{ch_i})
        end if;
    end for; route(MC_r^{ch_i}, ch_i); end for;
```

Our semantics changes the actual message and channel cardinalities, and "generates" messages by forwarding copies of the original one. Thereby the router keeps its read-only semantics. Listing 1.2 shows our semantics, in which all channel conditions are evaluated on a set of messages MC_r in parallel (cf. line 1). If a message content matches a channel condition, the message is added to a new message collection ($MC_r^{ch_i}$; cf. line 7) that is created per channel (cf. lines 4–6). Consequently, each channel condition results in a set of records (i.e., message collection). The default channel can either be added to the list of channels C beforehand or executed for all messages $m_r \in MC_r$ that are not in one of the message collections of one of the channels. For each channel ch_i the message collections $MC_r^{ch_i}$ can be routed in parallel (cf. line 9). Now, the message cardinality is $n{:}m$ and the channel cardinality changes to $1{:}n$, which means that several messages can be processed in one transaction.

Message Translator. The semantics of the translator is to translate one message content to another to make it understandable for its receiving processor or endpoint [11]. The translator has a channel and a message cardinality of 1:1. It does not generate new messages, but modifies the current one.

Our semantics changes the message cardinality to $n{:}n$, allowing for a more efficient translation of several messages in one MC_r. Again, the translation could be executed in parallel (e.g., by partitioning MC_r), depending on the underlying database system.

3   Relational Message Processing

In this section, we define the transactional process model for integration processes. Then we briefly reiterate the original processing semantics, before defining an adapted, but compatible processing semantics suitable for databases.

3.1   Transactional Process Model

The conventional integration process model $PM = (A, P, MC, CF, MF)$ denotes a directed acyclic process graph $PG = (N, E)$, consisting of nodes N and edges E. The nodes represent either a set of integration adapters A or message processors and (content-based) routers P, while the edges denote message channels with a control flow relation $CF \subseteq (A \cup P) \times P \times (A \cup P)$, and a message flow relation $MF \subseteq ((A \cup P) \times MC) \cup (MC \times (A \cup P))$. The actual data is wrapped in message collections MC. On a relational database, the sender and the receiver data is located in the database with integration logic (cf. case *(C1)*).

Furthermore, both applications store their data in the same database without shared schema access. The process instance is granted read-only access to the sender data and write access to the receiver data. Therefore we extended PM using database transactions, where the control and data flow are mapped to database constructs combined in a transactional PM (TPM). Figure 3 depicts

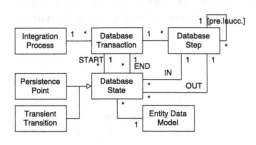

Fig. 3. Transactional process model

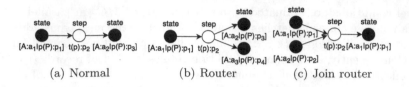

(a) Normal   (b) Router   (c) Join router

Fig. 4. Integration processing stereotypes.

the TPM, consisting of a process with arbitrarily many database transactions, while each transaction is composed of database steps (i. e., nodes N) and states (i. e., edges E). Database states can be persistent on disk or transient as data type definitions within one transaction. A transaction has at least a persistent start and end state and arbitrarily many intermediate persistent (e. g., table) or transient states (e. g., table type). A database step processes the data from a preceding state and moves it to a successor. The CF and MF process model relations map to the database states. The entity data models (i. e., message formats) are part of the state. Each state has a step "writing" to or "reading" from it. Transient states are parameter types within PL/SQL procedure calls. Figure 4 shows the described TPM in terms of the three most common stereotypes: normal (cf. Fig. 4(a)), 1:n routing (cf. Fig. 4(b)) and m:1 join routing (cf. Fig. 4(c)). Hereby, sending and receiving adapters or stateful processors represent persistent states (marked by black nodes) and message processors without state are denoted as steps with transient state (white nodes).

3.2 Interlude: Conventional Message Processing Semantics

For comparision, we recall the conventional message processing semantics [3], where a message is received by an EAI system from a sending application (i. e., persistent state: event sender or message queue) and processed within one transaction. This message is marked as "in-flight" in the sender, making it invisible for other consumers (e. g., process instance). When processed correctly, the resulting message is sent to the receiver application, the transaction is committed, and a notification is returned to the sender. In case of an exception, the intermediate message is cancelled and the transaction rolled back. The state in the sender changes to "stored", making it visible to its consumers again. The system reaches a consistent state and the message is redelivered from the sender. Consequently, the stereotypes in Fig. 4 require one transaction.

3.3 Database Message Processing

For database message processing, let us assume an instantiation of the transactional process model on a database. There, the message is processed in one transaction from persistence state to state (i. e., store and forward [13]). This might decompose the transactional, end-to-end processing into several sub-processes. In case of an error, the original message is redelivered from the last persistent state. However, on current database systems the transaction processing does not support setting a message or a collection MC of messages into "in-flight", but they are still visible to other transactions. Figure 5 illustrates the processing on a database denoting source and target tables as persistent states, and several transient message processors (i. e., no persistent state in-between). At the start of the processing, the messages msg_1 and msg_2 in the current message collection MC_{tx1} are read in transaction tx_1, and subsequently processed by transient message processors, before inserting the messages to the target table.

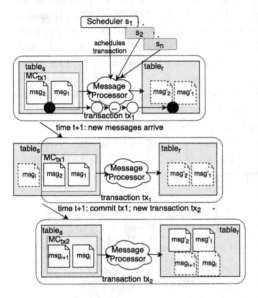

Fig. 5. Message Processing Model.

Since database triggers are executed within the sender's transaction context, we assume the transactions to be started by a controller or scheduler component (similar to the operator scheduling in streaming systems [6]). The scheduler processes a new transaction either on event, clocked or like a window-operator on a certain message collection size. Within one sub-process transaction, the data is read and deleted from the source table, processed and inserted into the target table, and then committed. Thereby, the messages in the source table $table_s$ remain visible for other consumers and become visible only in the target table $table_r$ after commit. In case of an exception, the messages will be redelivered in a new transaction (not shown). The difference compared to the conventional processing is that the messages in the source table remain visible, which can result in duplicate messages. To avoid duplicates, a stateful filter can be added that removes already processed messages. Similarly, the scheduler could ensure the "in-flight" transaction behavior by filtering the message collections.

While both mechanisms require additional, complex state management — possibly across distributed transactions, we decide for a third option. First, we separate integration processes by namespaces, and second we limit one scheduler to one transaction that starts only after a successful delivery (i. e., only one scheduler per process). While this prevents parallel processing, it ensures that message are processed consistently. Parallel processing can still be reached by increased collection sizes through partitioning. Within the database, the transactional processing is mandatory, making an automatic identification of the transactional boundaries important for our design.

3.4 Database Process Compilation with Transaction Identification

The message processing on a database might require more transactions than the conventional queue-based approach. Hence, for the evaluation of our approach, we built an integration to database process compiler that automatically determines transactions from a given process model, leveraging meta-data about the persistent states (e. g., aggregator, resequencer [11]). The corresponding persistent-set (p-set) algorithm is shown in Listing 1.3. The algorithm takes the process graph PG from Sect. 3.1 with adapters and message processors as nodes N and message channels as edges. We apply ECA rules [1] to all nodes in PG (cf. Listing 1.3). Thereby a rule r is defined as condition / action pair r : pset-match(IN : node $\in$ N, OUT : boolean), shown in Listing 1.4, and r : pset-execute(IN : node $\in$ N, OUT : transactions) in Listing 1.5. The pset-match is applied to all nodes in PG, while pset-execute is only applied to nodes for which pset-match evaluates to true. This is the case if the processor has a state — adapters are assumed to be persistent by default, and receive data through a channel. A transaction t is defined as structure with sets IN $\subseteq$ N for inbound states and OUT $\subseteq$ N for outbound states. The pset-execute creates a transaction for each identified state and adds all states to its IN and OUT sets, in a backward direction, to cover all stereotypical cases (cf. Fig. 4).

Listing 1.3. pset Algorithm

```
For all nodes node ∈ PG
  if r(pset)−match(node) then
    r(pset)−execute(node)
  end if;
```

Listing 1.4. pset match

```
r(pset)−match(node):
  return node.state=='persistent'
    && node.inDegree > 0
    && node.hasInChannel
```

Listing 1.5. pset execute

```
r(pset)−execute(node):
  t = new Transaction()
  t.OUT = node
  s = new Stack<Node>()
  for all inNode ∈ node.inNodes
    s.push(inNode)
  end for
  while (!s.empty)
    currentNode = s.pop()
    if currNode.state=='persistent'
      t.IN.add(currentNode)
    else
      t.add(currentNode)
    for all inNode' ∈ currNode.inNodes
      s.push(inNode')
    end for;
    end if;
  end while; return t
```

For instance, the processing stereotypes from Fig. 4 result in one transaction each. Furthermore, when applying the algorithm to a more complex process model with a router and join router pattern in Fig. 6(a), three transactions are identified. The scheduler executes these transactions ordered by their indices: $t_1 := (\text{IN}:se{:}1,s{:}1;\text{OUT}:s{:}2)$, $t_2 := (\text{IN}:s{:}2,s{:}3;\text{OUT}:ee{:}1)$, $t_3 := (\text{IN}:s{:}4,se{:}2,s{:}5;$ $\text{OUT}:ee{:}2)$. The process in Fig. 6(b) denotes a case that cannot be executed on a relational database with parallel transaction scheduling, since it would result in overlapping transactions identified by the pset algorithm. Consequently, this integration process would be rejected and the overlap highlighted: transactions t_1 and t_3 overlap in adapter se:2, and t_2 and t_3 in the stateful message

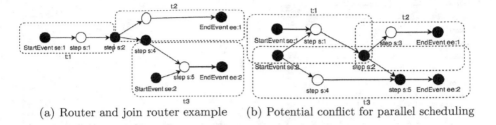

(a) Router and join router example   (b) Potential conflict for parallel scheduling

Fig. 6. Automatic detection of transactions for more complex integration processes.

processor s:2. Similar to the conventional message queuing, in our approach —
with ordered execution of transactions by only one scheduler — the transactions
can be executed in a consistent state.

4   Experimental Evaluation

In this section we evaluate the message throughput of the defined database
integration processes for distinct routing and transformation patterns, and their
latency in the context of the motivating scenario. The results implicitly illustrate
the semantic correctness of the redefined message processing for databases.

4.1   Benchmark Setup

We compare the message throughput of our approach deployed on a relational,
column-store database system (referred to as *SystemX*) with the open-source
integration system Apache Camel [12] (*Camel*) and our "table-centric" extensions of Camel with Datalog processing (*Camel-D*) from [16]. The integration
systems are running on an HP Z600 work station, equipped with two Intel X5650
processors clocked at 2.67 GHz with 12 cores, 24 GB of main memory, running a
64-bit Windows 7 SP1 and a JDK version 1.7.0 with 2 GB heap space.

The relational database system is running on the same machine. The benchmark definitions are taken from EIPBench
[18], which specifies benchmark configurations derived from "real-world" integration scenarios. EIPBench is a "data-aware"
micro-benchmark designed to measure the
throughput of messaging patterns. The message data sets are generated from the TPC-
H order to customer processing, but we only
generate messages based on orders. Table 1
summarizes the benchmark configurations
from [18] for the benchmark definitions that are relevant for our evaluation.

Table 1. EIPBench pattern benchmarks.

Benchmark	Description
CBR-A	simple cond.: OTOTALPRICE < 100.000
CBR-B	multiple conds.: OTOTALPRICE < 100.000, ORDERPRIORITY = "3-MEDIUM", OORDERDATE < 1970, OORDERSTATUS = "P"
MT-A	map names and filter entries according to a map. program

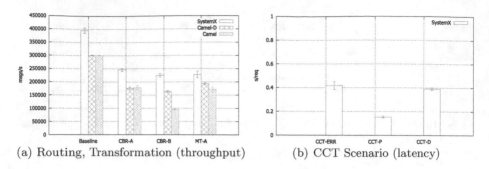

(a) Routing, Transformation (throughput)   (b) CCT Scenario (latency)

Fig. 7. Pattern throughput benchmark and process latencies for CCT scenario.

4.2 Pattern Throughput

For measuring the message throughput, we selected the CBR-A router and the MT-A transformation benchmarks [18], due to their similarity to the operations required for the CCT scenario. We added the CBR-B benchmark to further study the impact of more routing conditions. Figure 7(a) shows the benchmark results for our scheduled, store and forward processing (cf. Sect. 3.3). The database process reads the data from a persistent state — with MC_r size of 100, 000 (incl. all optional tables from Fig. 2) — and stores it into another in one transaction, after processing. See [16] for the Camel and Camel-D "micro-batching" extension. The baseline throughput for Camel routes and database processes (i. e., no intermediate processors) is added for comparison. For Camel and Camel-D, the baseline is the same, due to sharing the same pipeline engine. The results show a clear edge for the more data-centric, relational Camel-D processing for more complex routing and transformation patterns over the conventional Camel processing, while having similar results for the simpler CBR-A case. However, database processing outperforms both in all benchmarks. This is mainly due to the more efficient set-oriented data processing and the parallelization of read-only operations (e. g., used in the router). More routing conditions (cf. CBR-B) reduce the throughput for all three systems, however, have a lower impact on the more data-centric approaches (i. e., database processes, Camel-D).

4.3 Pattern Composition: Connected Car Telemetry (CCT)

The data sent from the vehicles in the motivating CCT example consists of approximately 304 B error code JSON messages (stored normalized in the column store) with fields like `"Trouble_Codes":"MIL is OFF0"` and approximately 762 B telemetry data with fields like `"Vehicle_Speed":"0km/h"` and `"Engine_Load":"18,8%"`. Fig. 7(b) shows that the latency depends on the path the messages take through the integration process. The routing of parking car data (*CCT-P*) shows the lowest latency, since parking cars are filtered early in the process. However, the telemetry data of driving cars (*CCT-D*) pass the filter and are then further processed in the same way as the error codes (*CCT-ERR*).

Consequently, the filter has no significant impact on the latency. Subsequently, the CCT-D and CCT-ERR data is enriched by the car owner's master data by lookup of the `Vehicle_ID_Number` and transformed into the receiver's format, showing a more significant reduction of the latency compared to the filter. This is due to the increasing amount of data and the more data-intensive operations.

5   Related Work

The aspect of moving EIP semantics and processes to the database (i.e., case *(C1)*) has been mentioned as future work by [2] in the context of process integration, however, was not addressed so far. We picked up the topic in our position paper [15], which discusses the expressiveness of database operations for integration processing on the content level. However, there is some work (mostly) in related domains, which we discuss subsequently.

Message Queuing: In the domain of declarative XML message processing, [4] defines a network of queues and XQuery data processors that are similar to our (persistent) states and transitions. This targets a subset of our approach (i.e., persistent message queuing; case *(C3)*), however, it does not cover *(C1)* integration semantics as relational database processes, which we target for analytic and business applications. In [7], a JMS-like message queuing engine is designed into the database, which allows for enqueuing and dequeuing messages, thus addressing case *(C2)*. This work is complementary to our approach, which uses the JMS endpoints as sources for EIP processing.

Business Processes: The work on executing business processes on the database, using an external process engine, evaluates nested views on the data and returns the results to the process engine (e.g., database engineering applications) [9]. More generally, [10] addresses the functional integration of data in federated database systems. Similar to our approach, data and control flow have been considered in business process management systems [21], which run the process engine optimally synchronized with the database. However, the work exclusively focuses on the optimization of workflow execution and does not consider application integration semantics on the database server level. In our work, we consider integration operations that are executed directly on the database, while no data is passed to a remote process engine or integration system.

Data-Intensive and Scientific Workflows: Based on the data patterns in workflow systems described by Russel et al. [20], modeling and data access approaches have been studied (e.g., by Reimann et al. [14]) in simulation workflows. The basic data management patterns in simulation workflows are ETL operations (e.g., format conversions, filters), a subset of the EIPs, and can be represented among others by our approach. The (map/reduce-style) data iteration pattern can be represented by scatter/gather or splitter/gather.

Data Integration: The data integration domain uses integration systems for querying remote data that is treated as local or "virtual" relations (e.g., Garlic

[8]) and evolved to relational logic programming, summarized by [5]. In contrast to remote queries, we extend the current EIP semantics in the database for EAI.

6 Discussion and Outlook

In this work, we defined a representation (cf. research question *Q1*) and processing semantics for application integration on relational databases (cf. *Q2*). We redefined 13 out of 38 message processing EIPs (e. g., implicitly: event message; point-to-point channel, channel adapter; explicitly: document message, datatype channel, content-based router, message filter, message translator, message store, canonical data model), discussed our new semantics and evaluated the approach for message throughput of distinct patterns as well as composition to our connected car telemetry scenario. Therefore, the integration processes are compiled to database processes by automatically identifying transactional contexts.

While the evaluation of database processes showed promising results for acceleration of the message throughput of more data-centric approaches (cf *Q3*), new patterns for non-functional aspects like exception handling [17,19] and message privacy [17] are currently only partially supported by database systems. Our approach allows for the compilation of integration processes to other database systems, but makes their configuration (e. g., through user-defined conditions or expressions) a task for database experts (i. e., not suitable for an integration developer or even business user), due to differences in SQL and PL/SQL constructs. Open research questions target the optimal message collection size for scenario workloads and parallel transaction processing that preserves the integration semantics. Moreover, the applicability to NoSQL databases is of interest.

Acknowledgments. We thank Dr Fredrik Nordvall Forsberg and Dr Norman May for valuable comments and discussions.

References

1. Act-Net Consortium, C.: The active database management system manifesto: a rulebase of ADBMS features. SIGMOD **25**(3), 40–49 (1996)
2. Batini, C., Lenzerini, M., Navathe, S.B.: A comparative analysis of methodologies for database schema integration. ACM Comput. Surv. **18**(4), 323–364 (1986)
3. Bernstein, P.A., Newcomer, E.: Principles of Transaction Processing. Morgan Kaufmann, Burlington (2009)
4. Böhm, A., Kanne, C.: Demaq/transscale: automated distribution and scalability for declarative applications. Inf. Syst. **36**(3), 565–578 (2011)
5. Cafarella, M.J., Halevy, A., Khoussainova, N.: Data integration for the relational web. VLDB **2**(1), 1090–1101 (2009)
6. Carney, D., Çetintemel, U., Rasin, A., Zdonik, S.B., Cherniack, M., Stonebraker, M.: Operator scheduling in a data stream manager. In: VLDB, pp. 838–849 (2003)
7. Gawlick, D., Mishra, S.: Information sharing with the oracle database. In: DEBS (2003)

8. Haas, L.M., Kossmann, D., Wimmers, E.L., Yang, J.: Optimizing queries across diverse data sources. In: VLDB, pp. 276–285 (1997)
9. Härder, T., Meyer-Wegener, K., Mitschang, B., Sikeler, A.: Prima - a DBMS prototype supporting engineering applications. In: VLDB (1987)
10. Hergula, K.: Daten- und Funktionsintegration durch föderierte Datenbanksysteme. Ph.D. thesis, University of Kaiserslautern (2003)
11. Hohpe, G., Woolf, B., Patterns, E.I.: Designing, Building, and Deploying Messaging Solutions. Addison-Wesley Longman, Boston (2003)
12. Ibsen, C., Anstey, J.: Camel in Action (2010)
13. Linthicum, D.S.: Enterprise Application Integration. Addison-Wesley, Boston (2000)
14. Reimann, P., Schwarz, H., et al.: Datenmanagementpatterns in simulationswork-flows. In: BTW, pp. 279–293 (2013)
15. Ritter, D.: What about database-centric enterprise application integration? In: ZEUS (2014)
16. Ritter, D.: Towards more data-aware application integration. In: Maneth, S. (ed.) BICOD 2015. LNCS, vol. 9147, pp. 16–28. Springer, Cham (2015). doi:10.1007/978-3-319-20424-6_3
17. Ritter, D., May, N., Rinderle-Ma, S.: Patterns for emerging application integration scenarios: a survey. Inf. Syst. **67**, 36–57 (2017)
18. Ritter, D., May, N., Sachs, K., Rinderle-Ma, S.: Benchmarking integration pattern implementations. In: DEBS, pp. 125–136 (2016)
19. Ritter, D., Sosulski, J.: Exception handling in message-based integration systems and modeling using BPMN. Int. J. Coop. Inf. Syst. **25**, 1–38 (2016)
20. Russell, N., Hofstede, A.H.M., Edmond, D., der Aalst, W.M.P.: Workflow data patterns: identification, representation and tool support. In: Delcambre, L., Kop, C., Mayr, H.C., Mylopoulos, J., Pastor, O. (eds.) ER 2005. LNCS, vol. 3716, pp. 353–368. Springer, Heidelberg (2005). doi:10.1007/11568322_23
21. Vrhovnik, M., Schwarz, H., Suhre, O., Mitschang, B., Markl, V., Maier, A., Kraft, T.: An approach to optimize data processing in business processes. In: VLDB, pp. 615–626 (2007)

Data Analysis and Data Mining

Preserving the Value of Large Scale Data Analytics over Time Through Selective Re-computation

Paolo Missier[✉], Jacek Cała, and Manisha Rathi

School of Computing Science, Newcastle University, Newcastle upon Tyne, UK
{Paolo.Missier,Jacek.Cala}@ncl.ac.uk, Manisha.Rathi@pwc.com

Abstract. A pervasive problem in Data Science is that the knowledge generated by possibly expensive analytics processes is subject to decay over time as the data and algorithms used to compute it change, and the external knowledge embodied by reference datasets evolves. Deciding when such knowledge outcomes should be refreshed, following a sequence of data change events, requires problem-specific functions to quantify their value and its decay over time, as well as models for estimating the cost of their re-computation. Challenging is the ambition to develop a decision support system for informing re-computation decisions over time that is both generic and customisable. With the help of a case study from genomics, in this paper we offer an initial formalisation of this problem, highlight research challenges, and outline a possible approach based on the analysis of metadata from a history of past computations.

Keywords: Selective re-computation · Incremental computation · Partial re-computation · Provenance · Metadata management

1 Your Data Will Not Stay Smart Forever

A general problem in Data Science is that the knowledge generated through large-scale data analytics tasks is subject to decay over time, following changes in both the underlying data used in their processing, and the evolution of the processes themselves. In this paper we outline our vision for a general system, which we refer to as ReComp, that is able to make informed re-computation decisions in reaction to any of these changes. We distinguish two complementary patterns, which we believe are representative of broad areas of data analytics.

1. Forwards ReComp. In this pattern, knowledge refresh decisions are triggered by changes that occur in the inputs to an analytics process, and are based on an assessment of the consequences of those changes on the current outcomes, in terms of expected value loss, or opportunities for value increase.

2. Backwards ReComp. Conversely, in this pattern the triggers are observations on the decay in the value of the outputs, and re-computation decisions are based on the expected value improvement following a refresh.

© Springer International Publishing AG 2017
A. Calì et al. (Eds.): BICOD 2017, LNCS 10365, pp. 65–77, 2017.
DOI: 10.1007/978-3-319-60795-5_6

In both cases, when a limited re-computation budget is available, estimates of the cost of refresh are needed. Cost may be expressed, for instance, as time and/or cost of cloud resource allocation.

To make these patterns concrete, we now present one instance of each.

1.1   Forwards: Impact Analysis

Data-intensive workflows are becoming common in experimental science. In genomics, for instance, it is becoming computationally feasible to process the human genome in search of mutations that may help diagnose a patient's genetic disease. In this scenario, which we expand on in Sect. 1.4, a diagnosis given in the past may be affected by improvements in the underlying genome sequencing technology, but also possibly in the bioinformatics algorithms, and by updates in the external reference data resources like the many human variation databases [4, 12]. In a Forwards ReComp scenario, each of these changes would trigger a decision process aimed at predicting which patients *would benefit the most* from a reassessment of their diagnosis. A limited budget leads to a problem of prioritising re-computations over a subset of the patients' population, using estimates of the future cost of re-enacting the workflows. A similar scenario occurs when long-running simulations are used e.g. to predict flood in large cities. In this case, the problem involves understanding the impact of changes to the urban topology and structure (new green areas, new buildings), without having to necessarily run the simulation anew every time.

1.2   Backwards: Cause Analysis

In machine learning, it is well-known that the predictive power of a trained supervised classifier tends to decay as the assumptions embodied by the data used for training are no longer valid. When new ground truth observations become available while using the model, these provide a measure of actual predictive performance and of its changes over time, i.e., relative to the expected theoretical performance (typically estimated a priori using cross-validation on the training set). We may view the trained model as the "knowledge outcome" and the problem of deciding when to refresh (re-train) the model as an instance of Backwards ReComp. Here the expected performance of the new model must be balanced against the cost of retraining, which is often dominated by the cost of generating a new training set.

1.3   The ReComp Vision

Figure 1 provides a summary of our vision of a `ReComp` meta-process for making recurring, selective re-computation decisions on a collection of underlying analytics processes for both these patterns. In both cases, the meta-process is triggered by observations of data changes in the environment (top). In the Forwards pattern, on the left, these are new versions of data used by the process.

This pattern requires the ability to (i) quantify the differences between two versions of a data, (ii) estimate the impact of those changes on a process outcomes, (iii) estimate the cost of re-computing a process, (iv) use those estimates to select past process instances that optimise the re-computation effort subject to a budget constraint, and (v) re-enact the selected process instances, entirely or partially.

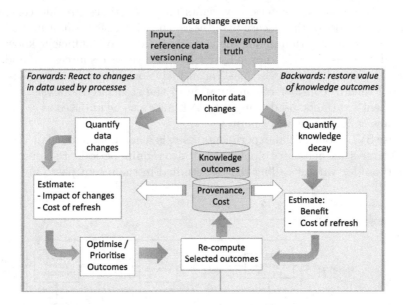

Fig. 1. Reference ReComp patterns

The Backwards pattern, on the right, is triggered by changes in data that can be used to assess the loss of value of knowledge outcomes over time, such as new ground truth data as mentioned earlier. This pattern requires the ability to (i) quantify the decay in the value of knowledge, expressed for instance in terms of model prediction power; (ii) estimate the cost and benefits of refreshing the outcome, and (iii) re-enact the associated processes.

To realise these patterns we envision a History database (centre). It contains both the outcomes that are subject to revision, and metadata about their provenance [19] and cost. Estimation models are learnt from the metadata which is updated upon each re-computation cycle. Note that for simplicity we only focus on changes in the data. Changes in the underlying processes, although relevant, require separate formalisation which is beyond the scope of this paper.

1.4   Example: Genetic Variants Analysis

The Simple Variant Interpretation (SVI) process [12] is designed to support clinical diagnosis of genetic diseases. First, patient *variants*, such as single-nucleotide polymorphisms, are identified by processing the patient's genome via a rese-quencing pipeline. Then, the SVI workflow (Fig. 2) takes these variants (about 25,000) and a set of terms that describe patient's *phenotype*, and tries to establish the deleteriousness of a small subset of those variants relevant to the phenotype by consulting external reference mutation databases. SVI uses the ClinVar[1] and OMIM Gene Map[2] reference databases, described in more detail later.

The reliability of the diagnosis depends on the content of those databases. Whilst the presence of deleterious variants may sometimes provide conclusive evidence in support of the disease hypothesis, the diagnosis is often inconclusive due to missing information about the variants, or due to insufficient knowledge in those databases. As this knowledge evolves and these resources are updated, there are opportunities to revisit past inconclusive or potentially erroneous diag-noses, and thus to consider re-computation of the associated analysis. Further-more, patient's variants, used as input to SVI, may also be updated as sequencing and variant calling technology improve.

We use SVI in our initial experiments, as it is a small-scale but fair represen-tative of large-scale genomics pipelines that also require periodic re-computation, such as those for variant calling that we studied in the recent past [3].

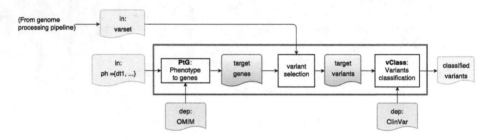

Fig. 2. The SVI workflow; inputs $\mathbf{x} = [varset, ph]$, external resources $\mathbf{D} = [OM, CV]$.

1.5 Contributions

We make the following contributions. (i) A semi-formal description of the selec-tive re-computation problem, which due to space constraints is limited to the *forwards* case, exemplified by the SVI case study; (ii) an outline of the associ-ated research challenges, and (iii) an initial analysis of the role of metadata, and specifically of provenance, as part of the ReComp vision.

This work reflects the initial phase of a project centred on the design of the ReComp meta-process (recomp.org.uk). What makes ReComp particularly chal-lenging is the ambition to develop a *generic* and customisable decision support system for informing data analytics re-computation decisions over time, in a setting where most approaches appear to be problem-specific.

---

[1] https://www.ncbi.nlm.nih.gov/clinvar.
[2] https://www.omim.org.

2   Reacting to Data Change Events

We formalise the *forwards* pattern of the ReComp problem in more detail, assuming an ideal scenario where a history of past program executions has been recorded, each data item is versioned, and a family of *data diff* functions, one for each of the data types involved in the computation, are available to quantify the extent of change between any two versions.

2.1   Definitions

Executions. Suppose we can observe a set of N executions of an analytics applications, which for simplicity we represent as single program P. Each execution $i : 1 \ldots N$ takes input x_i and may also use data from a set of *reference datasets* $D = \{D_1 \ldots D_m\}$ to produce value y_i. We assume that each of x_i and $D_j \in D$ may have multiple versions updated over time. We denote the version of x_i at time t as x_i^t, and the state of resource D_j at t as d_j^t. For each execution we also record its cost c_i^t (e.g. time or monetary expression that summarises the cost of cloud resources). We denote one execution of P that takes place at time t by:

$$\langle y_i^t, c_i^t \rangle = exec(P, x_i^t, d^t) \tag{1}$$

where $d^t = \{d_1^t \ldots d_m^t\}$ is the state at time t of each reference datasets D_j. As mentioned in Sect. 1.3, we assume that P stays constant throughout.

Example 1. SVI consists of one single process P, which initially is executed once for each new patient. It takes input pair $x = \langle varset, ph \rangle$ consisting of the set of patient's variants and patient's phenotype $ph = \{dt_1, dt_2, \ldots\}$ expressed using *disease terms* dt_i from the OMIM vocabulary, for example *Alzheimer's*. SVI associates a class label to each input variant depending on their estimated deleteriousness, using a simple "traffic light" notation:

$$\mathbf{y} = \{(v, class) | v \in varset, class \in \{\mathsf{red}, \mathsf{amber}, \mathsf{green}\}\}$$

$D = \{OM, CV\}$ consists of two reference databases, OMIM GeneMap and ClinVar, which are subject to periodic revisions. GeneMap maps human disorder terms dt to a set of genes that are known to be broadly involved in the disease:

$$OM = \{\langle dt, genes\ (dt) \rangle\}$$

Similarly, ClinVar maintains catalogue V of variants and associates a status to each variant $v \in V$, denoted $varst(v) \in \{\mathsf{unknown}, \mathsf{benign}, \mathsf{pathogenic}\}$:

$$CV = \{\langle v, varst(v) \rangle\}$$

SVI uses OM and CV to investigate a patient's disease (Fig. 2). Firstly, the terms in ph are used to determine the set of *target genes* that are relevant for the disease hypothesis. These are defined as the union of all the genes in $genes(dt)$ for each disease term $dt \in ph$. Then, a variant $v \in varset$ is selected if it is located on the *target genes*. Finally, the selected variants are classified according to their labels from $varst(v)$.   □

Data version changes. We write $x_i^t \to x_i^{t'}$ to denote that a new version of x_i has become available at time t', replacing the version x_i^t that was current at t. Similarly, $d_j^t \to d_j^{t'}$ denotes a new release of D_j at time t'.

Diff functions. We further assume that a family of type-specific *data diff* functions are defined that allow us to quantify the extent of changes. Specifically:

$$\textit{diff}_X(x_i^t, x_i^{t'}) \quad \textit{diff}_Y(y_i^t, y_i^{t'}) \tag{2}$$

compute the differences between two versions of x_i of type X, and two versions of y_i of type Y. Similarly, for each source D_j,

$$\textit{diff}_{D_j}(d_j^t, d_j^{t'}) \tag{3}$$

quantifies differences between two versions of D_j. The values computed by each of these functions are going to be type-specific data structures, and will also depend on how changes are made available. For instance, $d_j^t, d_j^{t'}$ may represent successive transactional updates to a relational database. More realistically in our analytics setting, and on a longer time frame, these will be two releases of D_j, which occur periodically. In both cases, $\textit{diff}_{D_j}(d_j^t, d_j^{t'})$ will contain three sets of added, removed, or updated records, respectively.

Example 2. Considering that the set of terms dt in OMIM is fairly stable, $\textit{diff}_{OM}(OM^t, OM^{t'})$ returns updates in their mappings to genes that have changed between the two versions (including possibly new mappings):

$$\textit{diff}_{OM}(OM^t, OM^{t'}) = \{\langle t, genes(dt)\rangle | genes(dt) \neq genes'(dt)\}$$

where $genes'(dt)$ is the new mapping for dt in $OM^{t'}$.

The difference between two versions of ClinVar consists of three sets: new, removed, and status-changed variants:

$$\textit{diff}_{CV}(CV^t, CV^{t'}) =$$

$$\{\langle v, varst(v) | varst(v) \neq varst'(v)\} \cup CV^{t'} \backslash CV^t \cup CV^t \backslash CV^{t'}$$

where $varst'(v)$ is the new class associated to v in $CV^{t'}$.   □

Change Impact. To describe the *impact* of a single change that occurs at time t' on an output y_i^t, $t < t'$, suppose we have computed new $y_i^{t'}$ using the new version of the data. E.g., if the change is $d_j^t \to d_j^{t'}$, we would have computed:

$$\langle y_i^{t'}, c_i^{t'} \rangle = exec(P, x_i^{t'}, d^{t'}) \tag{4}$$

where $d^{t'} = \{d_1^t \dots d_j^{t'} \dots d_m^t\}$. We define the impact of this change using type-specific function $f_Y()$ defined on the difference between the two versions of y_i:

$$imp(d_j^t \to d_j^{t'}, y_i^t) = f_Y(\textit{diff}_Y(y_i^t, y_i^{t'})) \in [0, 1] \tag{5}$$

where $y_i^{t'}$ is computed as in (4). In the case of our classified variants, for instance, $f_Y()$ could be defined as $f_Y(\textit{diff}_Y(y_i^t, y_i^{t'})) = 0$ if the diagnosis has not changed between two versions, and 1 if it has changed.

2.2   Problem Statement

Suppose a change is detected at t'; for simplicity let it be $d_j^t \rightarrow d_j^{t'}$ as above. Let $O^t = \{y_1^t, \ldots y_N^t\}$ denote the set of all outcomes that are current at time t. The ReComp goal is to select the optimal subset $O_{rc}^t \subseteq O^t$ of outcomes that, subject to budget C, maximise the overall impact of the change if they are re-computed:

$$\max_{O_{rc}^t \subset O^t} \sum_{y_i \in O_{rc}^t} imp(d_j^t \rightarrow d_j^{t'}, y_i^t), \quad \sum_{i:1}^{N} c_i^{t'} \leq C \tag{6}$$

As neither the impact nor the actual re-computation costs are known, however, solving this problem requires that we learn a set of cost and impact estimators:

$$\{\langle \widehat{imp}(d_j^t \rightarrow d_j^{t'}, y_i^t), \hat{c}_i^{t'} \rangle | y_i^t \in O^t\} \tag{7}$$

The optimisation problem can thus be written as:

$$\max_{O_{rc}^t \subset O^t} \sum_{y_i \in O_{rc}^t} \widehat{imp}(d_j^t \rightarrow d_j^{t'}, y_i^t), \quad \sum_{i:1}^{N} \hat{c}_i^{t'} \leq C \tag{8}$$

3   ReComp Challenges

A number of process and management challenges underpin this optimisation goal for the Forwards ReComp pattern.

3.1   Process Management Challenges

1. Optimisation of re-computation effort. Firstly, note that we must solve one instance of (8) for each data change event. Each instance can be formulated as the 0–1 knapsack problem in which we want to find vector $\mathbf{a} = [a_1 \ldots a_n] \in \{0,1\}^N$ that achieves

$$\max \sum_{i:1}^{N} v_i a_i \text{ subject to} \quad \sum_{i:1}^{N} w_i a_i \leq C \tag{9}$$

where $v_i = \widehat{imp}(d_j^t \rightarrow d_j^{t'}, y_i^t)$, $w_i = \hat{c}_i^{t'}$.

A further issue is whether multiple changes, i.e. to different data sources, should be considered together or separately. Also, in some cases it may be beneficial to group multiple changes to one resource, e.g. given $d_j^t \rightarrow d_j^{t'}$, we may react immediately or wait for the next change $d_j^{t'} \rightarrow d_j^{t''}$ and react to $d^t \rightarrow d^{t''}$.

2. Partial recomputation. When P is a *black box* process, it can only be re-executed entirely from the start. But a *white-box* P, like the SVI workflow, may benefit from the "smart re-run" techniques, such as those developed in the context of scientific data processing [1, 10].

Specifically, suppose that a granular description of P is available, in terms of a set of processing blocks $\{P_1 \ldots P_l\}$ where some P_j encodes a query to D_j. These, along with dataflow dependencies of the form: $P_i \rightarrow P_j$, form a directed workflow graph. If re-computation of P is deemed appropriate following a change in D_j, logically there is no need to restart the computation from the beginning, as long as it includes P_j (we know that a new execution of P_j will return an updated result). In theory, the exact minimal subgraph of P that must be recomputed is determined by the available intermediate data saved during prior runs [10]. An architecture for realising this idea is also presented in [9]. In practice, however, for data analytics tasks where intermediate data often outgrow the actual inputs by orders of magnitude, the cost of storing all intermediate results may be prohibitive. An open problem, partially addressed in [18], is to balance the choice of intermediate data to retain with the retention cost.

3. Learning cost estimators. This problem has been addressed in the recent past, but mainly for specific scenarios that are relevant to data analytics like workflow-based programming on clouds and grid, [11, 16]. But for instance [13] showed that runtime, especially in the case of machine learning algorithms, may depend on features that are specific to the input, and thus not easy to learn.

4. Process reproducibility issues. Actual re-computation of older processes P may not be straightforward, as it may require redeploying P on a new infrastructure and ensuring that the system and software dependencies are maintained correctly, or that the results obtained using new versions of third party libraries remain valid. Addressing these architectural issues is a research area of growing interest [2, 5, 17], but not a completely solved problem.

3.2   Data Management Challenges

5. Learning impact estimators. Addressing optimisation problem (8) requires that we first learn impact estimators (7). In turn, this needs estimating differences $\widehat{diff}_Y(y_i^t, y_i^{t'})$ for any $y_i^t \in O^t$ and any data change. But the estimators are going to be data- and change-specific and so, once again, difficult to generalise. This is a hard problem. In particular it involves estimating difference $diff_Y(y, y')$ between two values $y = f(x_1 \ldots x_k)$, $y' = f(x_1' \ldots x_k')$ for unknown function f, given changes to some of the x_i and the corresponding $diff_X(x_i, x_i')$. Clearly, some knowledge of function $f_Y()$ is required, which is also process-specific and thus hard to generalise into a reusable re-computation framework.

Example 3. Recalling our example binary impact function $f_Y()$ for CV, we would like to predict whether any new variant added to $CV^{t'}$ will change a patient's diagnosis. Using forms of provenance, some of which is described later (Sect. 4), we may hope not only to determine whether the variant is relevant for the patient, but also whether the new variant will change the diagnosis or it will merely reinforce it. This requires domain-specific rules, however, including checking whether other benign/deleterious variants are already known, and checking the status of an updated or new variant.   □

6. Proliferation of specialised *diff* functions. Suppose processes P_1 and P_2 retrieve different attributes from the same relational database D_j. Clearly, for each of them only changes to those attributes matter. Thus, data diff functions such as those defined in Sect. 2.1 are not only type-specific but also query-specific. For K processes and M resources, this potentially leads to the need for KM specialised *diff* functions.

7. Managing data changes. There are practical problems in managing multiple versions of large datasets, each of which may need to be preserved over time for potential future use by ReComp. Firstly, each resource will expose a different version release mechanism, standard version being the simple and lucky case. Once again, observing changes in data requires source-specific solutions. Secondly, the volume of data to be stored, multiplied by all the versions that might be needed for future re-computation, leads to prohibitively large storage requirements. Thus, providers' limitations in the versions they make available translates into a challenge for ReComp.

8. Metadata formats. ReComp needs to collect and store two main types of metadata: the detailed cost of past computations of P, which form ground truth data from which cost estimators can be learnt; and provenance metadata, as discussed next (Sect. 4). The former, although a simpler problem, requires the definition of a new format which, to the best of our knowledge, does not currently exist. Provenance, on the other hand, has been recorded using a number of formats, often system-specific. Even when the PROV model [14] is adopted, it can be used in different ways despite being designed to encourage interoperability. Our recent study [15] shows that the ProvONE extension to PROV (https://purl.dataone.org/provone-v1-dev) is a step forward to improve interoperability but it still is limited as it assumes that the traced processes are similar and implemented as a workflow.

3.3 The ReComp Meta-process

To address these challenges we have started to design a meta-process that can **observe executions** of the form (1), **detect and quantify data changes** using *diff* functions (2, 3), and **control re-computations** (4).

ReComp is an exercise in metadata collection and analysis. As suggested in Fig. 1, it relies on a history database that records details of all the elements that participate in each execution, as well as on the provenance of each output y_i, to provide the ground data from which estimators can hopefully be learnt. However, not all processes and runtime environments are *transparent* to observers, i.e., they may not allow for detailed collection of cost and provenance metadata. Thus, we make an initial broad distinction between *white-box* and *black-box* ReComp, depending on the level of detail at which past computations can be observed and the amount of control we have on performing partial or full re-computations on demand.

4   Provenance in White-Box Recomp

As an example of the role of metadata, we analyse how provenance might be used in a *white-box*, fully transparent ReComp setting. Our goal is to reduce the size of the optimisation problem, that is, to identify those $y^t \in O^t$ that are *out of scope* relative to a certain data change. Formally, we want to determine the outputs $y_i^t \in O^t$ such that for change $d_j^t \to d_j^{t'}$, we can determine that

$$imp(d_j^t \to d_j^{t'}, y_i^t) = 0$$

For example, the scope of a change in ClinVar that reflects a newly discovered pathogenic status of a variant can be safely restricted to the set of patients who include that variant and whose phenotype renders the associated gene.

To achieve such filtering in a generic way, suppose we have access to the provenance of each y_i^t. While this refers generally to the history of data derivations from inputs to outputs through the steps of a process [19], in our setting we are only interested in recording which data items from D_j were used by P during execution. In a favourable yet common scenario, suppose that D_j consists of a set of records, and that P interacts with D_j through well defined queries Q_{D_j} using for instance a SQL or a keyword search interface. Then, the provenance of y_i^t includes all the data returned by execution of each of those queries: $Q_{d_j^t}$, along with the derivation relationships (possibly indirect) from those to y_i^t. Instead, here we take an *intensional* approach and record the queries themselves as part of the provenance:

$$prov(y_i^t) = \{Q_{D_j}, j : i \dots m\}$$

where each query is specific to the execution that computed y_i^t. The rationale for this is that, by definition, an output y_i^t is in the scope of a change to d_j if and only if P used any of the records in $diff_{D_j}(d_j^t, d_j^{t'})$, that is, if and only if Q_{D_j} returns a non-empty result when executed on the *difference* $diff_{D_j}(d_j^t, d_j^{t'})$.

In practice, when D_j is a set of records, we may naturally also describe $diff_{D_j}(d_j^t, d_j^{t'})$ as comprising of three sets of records – new: $r \in d_j^{t'} \backslash d_j^t$, removed: $r \in d_j^t \backslash d_j^{t'}$, and updated: $r \in d_j^{t'} \cap d_j^t$ where some value has changed. This makes querying the differences a realistic goal, requiring minor adjustments to Q_{D_j} (to account for differences in format), i.e., we can assume we can execute $Q_{D_j}(d_j^{t'} \backslash d_j^t)$, $Q_{D_j}(d_j^t \backslash d_j^{t'})$, and $Q_{D_j}(d_j^{t'} \cap d_j^t)$.

Example 4. Consider patient Alice whose phenotype is *Alzheimer's*. For SVI, this is also the keyword in query to GeneMap: Q_{OM} = "Alzheimer's". Suppose that performing the query at time t returns just one gene: $Q_{OM}(om^t)$ = {PSEN2}. Then, SVI uses that gene to query CV, and suppose that nothing is known about the variants on this gene: $Q_{CV}(cv^t) = \emptyset$. At this point, the provenance of SVI's execution for Alice consists of the queries: $\{Q_{OM} \equiv$ "Alzheimer's", $Q_{CV} \equiv$ "PSEN2"$\}$. Suppose that later at time t' CV is updated to include just one new deleterious variant along with the gene it is situated on: $\langle 227083249, \texttt{PSEN2}, \texttt{pathogenic} \rangle$. When we compute $\textit{diff}_{CV}(cv^t, cv^{t'})$, this tuple is included in $cv^{t'} \setminus cv^t$ and is therefore returned by a new query Q_{CV} on this difference set, indicating that Alice is in the scope of the change. In contrast, executing on the same diff set a similar CV query from another patient's provenance, where PSEN2 is not a parameter, returns the empty set signaling that the patient is definitely not affected by the change.   □

Note that similar idea of using provenance for partial re-computation has been studied before [7,8]. The goal was to determine precisely the fragment of a data-intensive program that needed to be re-executed to *refresh* stale results. However, it required full knowledge of the specific queries, which we do not. A formal definition of correctness and minimality of a provenance trace with respect to a data-oriented workflow is proposed by members of the same group [6]. The notion of *logical provenance* may be useful in our context, too; once mapped to PROV that has since emerged as a standard for representing provenance.

Note also, that the technique just sketched can only go as far as narrowing the scope of a change, yet reveals little about its impact. Still, in some cases we may be able to formulate simple domain-specific rules for qualitative impact estimation that reflect our propensity to accept or prevent false negatives, i.e. ignoring a change that has an impact. An example of such a conservative rule would be "if the change involves a new *deleterious* variant, then re-compute all patients who are in the scope of that change".

The earlier example illustrates how queries saved from previous executions can be used to determine the scope of a change, assuming that the queries themselves remain constant. However this assumption can be easily violated, also in our running example. Suppose that at time t OM is updated instead of CV, e.g. an additional gene X is related to Alzheimer's. We now have $Q_{OM}(om^{t'}) = \{\texttt{PSEN2}, \texttt{X}\}$, therefore $Q_{CV} \equiv$ "PSEN2, X" rather than just "PSEN2" as recorded in the provenance. This brings the additional complication that the stored queries may need to be updated prior to being re-executed on the diff records.

Finally, note that in this specific example, when the change occurs in the input, that is, in the patient's genome, the scope of the change consists of just one patient. In this case, it may well be beneficial to always re-compute, as computing $\textit{diff}_X(x_i^t, x_i^{t'})$ to determine which parts of the genome have changed and whether the change will have any impact may be just as expensive, and thus inefficient. These questions are the subject of our current experimentation.

4.1 Conclusions

In this paper we have made the case for a new strand of research to investigate strategies for the selective, recurring re-computation of data-intensive analytics processes when the knowledge they generate is liable to decay over time. Two complementary patterns are relevant: *forwards impact analysis*, and *backwards cause analysis*. With the help of a case study in genomics we offered a simple formalisation of the former,[3] and outlined a number of challenges in designing a generic framework for a broad family of underlying analytics processes.

To address these problems, we propose a `ReComp` meta-process that can collect metadata (cost, provenance) on a history of past computation and use it to learn cost and impact estimators, as well as to drive partial re-computation on a subset of prior outcomes. As an example of our early investigation in this direction, we have discussed the role of data provenance in an ideal white-box scenario.

References

1. Altintas, I., Barney, O., Jaeger-Frank, E.: Provenance collection support in the kepler scientific workflow system. In: Moreau, L., Foster, I. (eds.) IPAW 2006. LNCS, vol. 4145, pp. 118–132. Springer, Heidelberg (2006). doi:10.1007/11890850_14
2. Burgess, L.C., Crotty, D., de Roure, D., Gibbons, J., Goble, C., Missier, P., Mortier, R., Nichols, T.E., O'Beirne, R.: Alan Turing Intitute Symposium on Reproducibility for Data-Intensive Research - Final Report (2016)
3. Cała, J., Marei, E., Xu, Y., Takeda, K., Missier, P.: Scalable and efficient whole-exome data processing using workflows on the cloud. Future Gener. Comput. Syst. **65**(Special Issue: Big Data in the Cloud), 153–168 (2016)
4. Cooper, G.M., Shendure, J.: Needles in stacks of needles: finding disease-causal variants in a wealth of genomic data. Nat. Rev. Genet. **12**(9), 628–640 (2011)
5. Freire, J., Fuhr, N., Rauber, A.: Reproducibility of data-oriented experiments in e-science (Dagstuhl Seminar 16041). Dagstuhl Reports **6**(1), 108–159 (2016)
6. Ikeda, R., Das Sarma, A., Widom, J.: Logical provenance in data-oriented workflows? In: 2013 IEEE 29th International Conference on Data Engineering (ICDE), pp. 877–888. IEEE, April 2013
7. Ikeda, R., Salihoglu, S., Widom, J.: Provenance-based refresh in data-oriented workflows. In: Proceedings of the 20th ACM International Conference on Information and Knowledge Management, pp. 1659–1668 (2011)
8. Ikeda, R., Widom, J.: Panda: a system for provenance and data. In: Proceedings of the 2nd USENIX Workshop on the Theory and Practice of Provenance (TaPP 2010), vol. 33, pp. 1–8 (2010)
9. Koop, D., Santos, E., Bauer, B., Troyer, M., Freire, J., Silva, C.T.: Bridging workflow and data provenance using strong links. In: Gertz, M., Ludäscher, B. (eds.) SSDBM 2010. LNCS, vol. 6187, pp. 397–415. Springer, Heidelberg (2010). doi:10.1007/978-3-642-13818-8_28
10. Ludäscher, B., Altintas, I., Berkley, C., Higgins, D., Jaeger, E., Jones, M., Lee, E.A., Tao, J., Zhao, Y.: Scientific workflow management and the Kepler system. Concurrency Comput. Pract. Exp. **18**(10), 1039–1065 (2006)
11. Malik, M.J., Fahringer, T., Prodan, R.: Execution time prediction for grid infrastructures based on runtime provenance data. In: Proceedings of WORKS 2013, pp. 48–57, New York, USA. ACM Press (2013)
12. Missier, P., Wijaya, E., Kirby, R., Keogh, M.: SVI: a simple single-nucleotide human variant interpretation tool for clinical use. In: Ashish, N., Ambite, J.-L. (eds.) DILS 2015. LNCS, vol. 9162, pp. 180–194. Springer, Cham (2015). doi:10.1007/978-3-319-21843-4_14

---

[3] Analysis of the specific "backwards" cases will appear in a separate contribution.

13. Miu, T., Missier, P.: Predicting the execution time of workflow activities based on their input features. In: Taylor, I., Montagnat, J., (eds.) Proceedings of WORKS 2012, Salt Lake City, US. ACM (2012)
14. Moreau, L., Missier, P., Belhajjame, K., B'Far, R., Cheney, J.T.: PROV-DM: the PROV data model. Technical report, World Wide Web Consortium (2012)
15. Oliveira, W., Missier, P., Ocaña, K., Oliveira, D., Braganholo, V.: Analyzing provenance across heterogeneous provenance graphs. In: Mattoso, M., Glavic, B. (eds.) IPAW 2016. LNCS, vol. 9672, pp. 57–70. Springer, Cham (2016). doi:10.1007/978-3-319-40593-3_5
16. Pietri, I., Juve, G., Deelman, E., Sakellariou, R.: A performance model to estimate execution time of scientific workflows on the cloud. In: Proceedings of WORKS 2014, pp. 11–19. IEEE, November 2014
17. Stodden, V., Leisch, F., Peng, R.D.: Implementing Reproducible Research. CRC Press, Boca Raton (2014)
18. Woodman, S., Hiden, H., Watson, P.: Workflow provenance: an analysis of long term storage costs. In: Proceedings of WORKS 2015, pp. 9: 1–9: 9 (2015)
19. PROV-Overview. An Overview of the PROV Family of Documents, April 2013

Statistical Data Analysis of Culture for Innovation Using an Open Data Set from the Australian Public Service

Warit Wipulanusat$^{(\boxtimes)}$, Kriengsak Panuwatwanich,
and Rodney A. Stewart

Griffith University, Gold Coast, Australia
warit.wipulanusat@griffithuni.edu.au

Abstract. Opportunities for replicating large data sets that have already been collected by government, and made available to the public to provide the possibility of statistical data analysis, are starting to emerge. This study examines the factor structure of ambidextrous culture for innovation. Survey data was extracted from the State of the Service Employee Census 2014 comprising 3,125 engineering professionals in Commonwealth of Australia departments. Data were analysed using exploratory factor analysis (EFA) and confirmatory factor analysis (CFA). EFA returned a two-factor structure explaining 61.1% of the variance of the construct. CFA revealed that a two-factor structure was indicated as a validated model (GFI = 0.99, AGFI = 0.98, RMSEA = 0.05, RMR = 0.02, IFI = 0.99, NFI = 0.99, CFI = 0.99, and TLI = 0.98). From the results, the two factors extracted as characterising ambidextrous culture for innovation were innovative culture and performance-oriented culture.

Keywords: Open data set · Culture for innovation · Exploratory factor analysis · Confirmatory factor analysis

1 Introduction

An effective government culture is one that focuses on shared norms and basic values pertaining to innovation in the work environment and reflecting innovative practices, procedures, policies and structures. In this study, the dimensionality of culture for innovation was analysed using an exploratory factor analysis (EFA) and confirmatory factor analysis (CFA). The inter-relationship between variables and the factor structure of their measure was analysed using EFA. Subsequently, CFA was used to test the fit of the model.

Appropriate variables can be identified using EFA. Additionally, the relationships among large numbers of variables in the most general form can be analysed using this method through explanations in terms of their common underlying dimensions [1]. A number of factors were retained in the construct based on results from the EFA together with a clear estimation of the factor structures to be used as the measures of this construct. To confirm the validity of the measurement scale, CFA was sequentially conducted. CFA is a theory-driven technique which tests the hypotheses for a factor

© Springer International Publishing AG 2017
A. Calì et al. (Eds.): BICOD 2017, LNCS 10365, pp. 78–89, 2017.
DOI: 10.1007/978-3-319-60795-5_7

structure. This determines the validity of theoretical structures by testing the causal links among variables [2, 3]. This served to strengthen the results of the EFA and support the recognised factor structures derived from the EFA process. Researchers can use CFA to evaluate the structure of factors, identify the dimensions of a construct, and consider whether particular patterns of loadings match the data [1].

2 Literature Review

2.1 Ambidextrous Culture for Innovation

The ambidextrous culture for innovation is classified into two sets of cultural features. The first type of organisational culture is an innovative culture where an organisation orients toward experimenting with new solutions by exploring new resources, breaking through existing norms, and valuing flexibility, adaptability, creativity, risk taking, and entrepreneurship. Such an innovative culture encourages employees to implement new services, new technologies for product development, new organisational routines and structures [4]. Cultivating employee orientation toward innovation may lead them to feel that the organisation is full of spirit, can manage uncertainty in the work environment, and reduces unfavourable consequences in the organisation. Innovation orientation is also likely to lead to truly innovative breakthroughs due to its emphasis on creativity. Lægreid et al. [5] assert that innovative cultures can ensure employees consistently perceive an innovative cultural orientation, thus serving as guidance when they face challenges affecting innovation outcomes.

The second type of ambidextrous culture for innovation stresses organisational performance orientation, in which organisational characteristics are categorised into three elements: (a) strongly developed goal orientation; (b) a focus on task performance; and (c) a strong emphasis on quality of service delivery. Such a performance-oriented culture is accepted to increase innovation in public sectors because governments need to respond to the demands of clients and citizens to achieve targets and performance evaluation processes. According to Lægreid et al. [5], public sectors which have strongly developed performance-oriented cultures are more likely to have an innovative culture and promote innovative activities compared to other public sectors.

2.2 Open Data Set from the Australian Public Service

Opportunities for replicating large data sets in public management studies are starting to emerge. An open data set is defined as data that has already been collected by public or private organisations, and made available to the public to provide the possibility of study replications [6]. There are many advantages of using open data sets. One major advantage is the extent of the data available from studies which are conducted on a large national scale and cover a broad geographic scope. In contrast, a researcher is likely to collect a smaller sample than an organisation would. The second advantage is the professional data collection skills of research experts in large organisations may not be readily available to smaller studies [7]. The third advantage comes in terms of expense and time savings. Collecting data for a national survey with a sample size of

1,500 to 2,000 can be very expensive with the cost potentially exceeding $300,000 in most cases [8]. The design and distribution of the questionnaire, as well as collecting and inputting the data, can take a considerable amount of money and time. Finally, when data is collected on a national scale, the large sample size offered by national databases has a higher likelihood of representation of the population and can furnish the power needed to effect generalisations of the findings. This is true in the context of this research, where the open data sets provided the above advantages to examine innovation in public sectors.

However, while the advantages of using open data sets are very practical, the disadvantages have a tendency to be more conceptual. A major disadvantage is that the data has not been designed particularly to address the researchers own questions. Researchers are not involved in the creation of the questionnaire, the types of questions, or the measurement scales; therefore, the researcher is limited to only using the available data. The second disadvantage is that the data collection process was not administered by researchers who opt to use open data. As a result, identification of any bias encountered during the collection process may not be detected as the survey instrument was designed and tested by others. The third disadvantage consists of access to the data, as there may be restrictions and conditions.

Governments in industrialised countries are now surveying their employees to consider their views and attitudes about the workplace, management, and human resources. In Australia, the State of the Service Employee Census examines the views of APS employees on workplace issues such as leadership, learning and development, health and wellbeing, and job satisfaction. The large sample size surveyed in the State of the Service Employee Census and its widespread coverage of federal departments can be applied for the generalisability of federal bureaucracy results on a national scale, including the perspectives of engineering professions because it also reaches the desired population of engineers within the federal government. In this study the potential limitations of using open data sets do not apply for the State of the Service Employee Census. The data set is in the Australian public service domain and can be accessed to address the research questions in this study. Using the census data assures no eligible respondents have been excluded from the survey sample, removing sampling bias and reducing sample error. There were no restrictions to use the data set and it is compatible with SPSS with only basic data cleaning required. Therefore, the most recent State of the Service Employee Census 2014 was adopted to conduct the quantitative analysis.

From an Australian public sector perspective, Torugsa and Arundel [9] employed a sample of 4,369 federal employees from the State of the Service Employee Census 2011, to explore factors associated with complexity and examined how complexity affected innovation outcomes. Torugsa and Arundel [10] also used the census data to investigate the characteristics of innovation implemented at the workgroup level and evaluated the different dimensions of the most significant innovation in the work group. Thus, open data sets have been identified by researchers to be valuable sources of information to examine innovation in public sectors.

3   Methodology

Data released by the Australian Public Service Commission (APSC) from the 2014 APS employee census was used for this study. This survey instrument was administered to gather data from civil servants in Commonwealth departments using an online survey. The State of the Service Employee Census 2014 targeted full-time, permanent civil servants of federal departments throughout the Federal Government and was collected from 12 May to 13 June 2014. The survey was distributed to a random sample of employees from federal departments with 100 or more employees throughout Australia. The survey was self-administered on a voluntary basis. Employees were invited to participate in the survey using email solicitation and a web-based questionnaire. An e-mail was sent to the participants with a link to the questionnaire, so respondents could access the survey at a time and location convenient to their situation.

This study draws on the perspectives of engineers as key informants. Of all respondents, 3,570 reported their type of work as Engineering and Technical Family. Cases where an entire section was left blank were eliminated from the sample as nonresponsive, leaving a total of 3,125 observations available for this study. The survey items were measured on a 5 point Likert scale (Strongly Disagree = 1 and Strongly Agree = 5).

The quantitative analysis commenced with multivariate statistics to analyse the data using the Statistical Package for the Social Sciences (SPSS) version 22 software. To clean the data and decrease systematic errors, missing values, outliers, and distribution of all measured variables were examined. This data screening was conducted to ensure there would be no corrupted data which could affect the accuracy of the estimation in the subsequent analysis [11].

4   Results

4.1   Exploratory Factor Analysis

To estimate the validity of the scales, exploratory factor analysis (EFA) was conducted to reduce the large number of variables into a smaller, more controllable set of dimensions [1, 2] and to summarise the data prior to including them into confirmatory factor analysis. The EFA is a multivariate statistical technique used to search for structure among a set of observed variables (i.e. questions in the questionnaire) and subsequently group highly inter-correlated observed variables into a latent factor. This approach allows researchers to be scientifically confident that the observed variables reproduce one single variable as a latent variable. Generally, the EFA can be performed by employing either R-type or Q-type factor analysis. The former identifies a set of dimensions that is latent in a large set of items, while the latter condenses large numbers of people into distinctly separate groups [1]. The R-type was adopted in this study because the main purpose of this stage was to allocate the observed variables into a set of new composite dimensions (factors).

According to Russell [12], the EFA must be conducted to determine whether a theoretical construct is a single or multidimensional factor, and to give a clear estimation of the factor structure of the measures. As a preliminary analysis, the EFA is very useful in the absence of an adequately detailed theory about the relations of variables to the underlying constructs [2]. Although all measured variables in the construct were derived from a comprehensive literature review and prior studies, the EFA was deemed necessary because these variables had not been operationalised extensively within the public sector context. Therefore, given that the model construct was measured by an independent scale, the EFA was conducted separately for the model construct to see if the chosen variables loaded on the expected latent factors were adequately correlated and, met the criteria of reliability and validity within this sample.

In order to assess the factorability of the data and ensure sampling adequacy, Bartlett's test of sphericity and the Kaiser-Meyer-Olkin (KMO) measure of sampling adequacy were applied. According to Tabachnick and Fidell [11], data is factorable when the KMO is above the minimum acceptable level of 0.60. KMO values over 0.8 indicate that included variables are 'meritoriously' predicted without error by other variables. The KMO value of the variables was 0.896, which indicated sampling adequacy such that the values in the matrix were sufficiently distributed to conduct factor analysis. The value obtained by Bartlett's test of sphericity, $\chi^2(55)$ was 16804.95, which was highly significant at $p < 0.001$ level, indicating that the data were approximately multivariate normal [13].

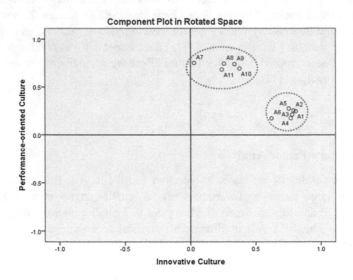

Fig. 1. A geometrical representation of the PAO construct

A principal component analysis (PCA) with varimax rotation was initially conducted to examine the validity of the construct. The goal of PCA is to extract the maximum variance from the data set with each component. A total of eleven variables were selected to operationally define the ACI construct. The presence of two components with eigenvalues greater than 1 was indicated by the initial principal component analysis. A geometrical approach can be utilised by the EFA where factors in a coordinate system can be visualized by variables plotted on the axes of a graph [14]. When the coordinates of variables are in close proximity to each graph, this represents the strength of the relationship between that variable and each factor. The variables were plotted as a function of the factors, as shown in Fig. 1. Six variables (A1, A2, A3, ..., A6) have high factor loadings (i.e., a strong relationship) with factor 1 (innovative culture: horizontal axis) but have a low correlation with factor 2 (performance-oriented culture: vertical axis). In comparison, five variables have strong relationships with performance-oriented culture but low correlation with innovative culture. ·

The Catell's scree test employs a graphical plot of the eigenvalue of the factor in their order of extraction in which a sudden change of slope in the graph indicates the maximum number of factors to be extracted and determines the number of factors to retain. A horizontal and a vertical line starting at each end of the curve were inserted to determine whether an abrupt change of slope had occurred. Examination of the scree plot indicated that a sudden change of slope occurred after the second component (See Fig. 2).

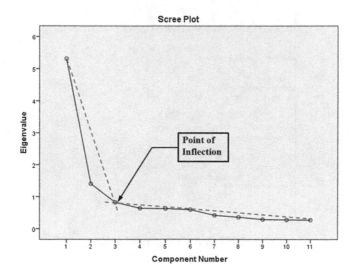

Fig. 2. Scree plot of principal component of the PAO construct

Prior to extracting factors, communality estimates must be generated. Communality is the proportion of observed variance accounted for by the common factors. These values represent the total amount of variance for an item explained by the extracted factors. The communality is denoted by h^2 and is the summation of the squared factor loadings of a variable across factor [11]. Generally, a variable is excluded from the analysis if it has low communalities (less than 0.20), which means that 80% is unique variance. This is because the objective of factor analysis is to describe the variance through the common factors. The formula for deriving the communalities is:

$$h_j^2 = a_{j1}^2 + a_{j2}^2 \cdots\cdots + a_{jm}^2 \tag{1}$$

Where a equals the loadings for j variables.

Parallel analysis was also conducted for factor extraction, in which the eigenvalues derived from actual data were compared with the eigenvalues resulting from the random data. Factors are retained when actual eigenvalues exceed random data. SPSS syntax, as depicted in Fig. 3, (adapted from O' Connor [15]) can be used to achieve this by simply specifying the number of cases, variables, data sets, and the desired percentile for the analysis at the beginning of the program.

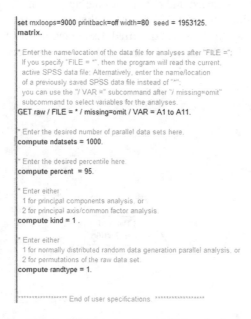

```
set mxloops=9000 printback=off width=80  seed = 1953125.
matrix.

* Enter the name/location of the data file for analyses after "FILE =";
  If you specify "FILE = *", then the program will read the current,
  active SPSS data file; Alternatively, enter the name/location
  of a previously saved SPSS data file instead of "*";
  you can use the "/ VAR =" subcommand after "/ missing=omit"
  subcommand to select variables for the analyses.
GET raw / FILE = * / missing=omit / VAR = A1 to A11.

* Enter the desired number of parallel data sets here.
compute ndatsets = 1000.

* Enter the desired percentile here.
compute percent  = 95.

* Enter either
  1 for principal components analysis, or
  2 for principal axis/common factor analysis.
compute kind = 1 .

* Enter either
  1 for normally distributed random data generation parallel analysis, or
  2 for permutations of the raw data set.
compute randtype = 1.

***************** End of user specifications. *****************
```

Fig. 3. SPSS syntax for parallel analysis

The results, as shown in Table 1, indicate that only the first 2 actual eigenvalues are greater than those generated by random data, and therefore, these would be retained.

During the operationalisation, the structure of the data for the ACI construct was classified into two factors: innovative culture and performance-oriented culture.

Table 1. Parallel analysis

Component number	Eigenvalues derived from actual data	Eigenvalues derived from random data	Decision
1	5.315	1.094	Accept
2	1.404	1.068	Accept
3	0.823	1.049	Reject
4	0.637	1.032	Reject

To date, no study has developed a framework to provide measurement variables for these two factors. The EFA was employed to empirically examine if the variables for the ACI construct could indeed be represented by these two factors. The initial principal component analysis indicated the presence of two components with eigenvalues greater than 1, which explained 61.1% of the total variance.

Using the factor loadings in Table 2, the communality of variable A1 was calculated using the aforementioned formula:

$$h_{A1}^2 = 0.806^2 + 0.248^2 = 0.712 \tag{2}$$

Communalities ranged from 0.415 to 0.712, indicating relatively strong data. The patterns for the rotated factor loading, presented in Table 2, revealed that all variables were well above the 0.50 threshold level. Moreover, the patterns for the loading of the variables onto their respective factors were also consistent with the operational definition of ACI construct in the literature review, assuring its robustness for this scale development. As with the original operationalisation, the ACI construct consisted of the following two factors:

- ACI1: Innovative culture (6 variables, component 1); and
- ACI2: Performance-oriented culture (5 variables, component 2).

Table 2. Rotated factor loadings of the ACI construct

Variable: Description	Component		h^2
	1	2	
A1: My agency prioritises ideas development.	0.81	0.25	0.71
A2: Most managers encourage innovation.	0.79	0.26	0.69
A3: Internal communication is effective.	0.78	0.21	0.65
A4: Change is managed well in my agency.	0.77	0.18	0.62
A5: My agency emphasises innovation.	0.75	0.28	0.64
A6: My workplace provides access to learning.	0.62	0.17	0.42
A7: Managers make sure procedure is follow.	0.03	0.75	0.57
A8: My agency emphasises task delivery.	0.26	0.74	0.62
A9: Most managers ensure their team delivers.	0.34	0.74	0.66
A10: My agency prioritises goals achievement.	0.38	0.69	0.62
A11: My agency emphasises standardised services.	0.24	0.68	0.52

4.2   Confirmatory Factor Analysis

Confirmatory factor analysis (CFA) is a method which enables evaluations of how well these measured variables represent the latent constructs. Although similar to exploratory factor analysis (EFA), but CFA differs in the sense that the number of constructs and indicators need to be specified before the computation of results. As such, it is considered as a "confirmatory" method [16]. CFA was conducted to examine whether these dimensions contributed to an overall construct of ambidextrous culture for innovation. To analyse the data using CFA, the Analysis of Moment Structures (AMOS) version 22 was employed to allow the data from an SPSS analysis set to be directly used in the AMOS calculation [17]. A 10-item measure of the two ACI dimensions derived from EFA was tested in AMOS. The results of the CFA conducted on AMOS have been presented in Fig. 4. For the model to be considered as having an acceptable fit, all eight indices were measured against the criteria: GFI, AGFI, IFI, NFI, CFI and TLI > 0.90; RMSEA < 0.08; and RMR < 0.05 [1]. The initial fit indices were unsatisfactory. Examination of the modification indices provided clear evidence of misspecified error covariance associated with variables A3, A4, A7, and A9, thus they were removed from the model. The likelihood ratio test revealed yielded a χ^2 of 129.12 and df = 13. The respecified model showed evidence of satisfactory model fit: GFI = 0.99, AGFI = 0.98, RMSEA = 0.05, RMR = 0.02, IFI = 0.99, NFI = 0.99, CFI = 0.99, and TLI = 0.98.

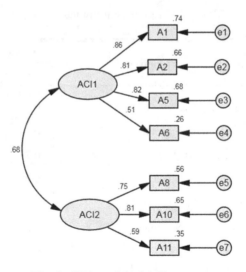

Fig. 4. CFA model of ACI construct

All of the variables loaded significantly ($p < 0.001$) on their respective constructs, as evidenced by the loading being greater than the threshold level of 0.50 (ranging from 0.51 to 0.86), indicating convergent validity. Almost all the variables had R^2 values greater than 0.50. The variables with low R^2 values were retained since their loadings were substantial and highly significant. The correlation coefficient between the two factors was 0.68. Thus, being less than 0.85, discriminant validity was confirmed.

Composite reliability (CR) measures the internal reliability of all the variables in their measurement of a construct. An average variance extracted (AVE) determines the amount of variance in the measured indicators captured by the latent construct. Bagozzi and Yi [18] suggested 0.60 and 0.50 as the minimum value of composite reliability and average variance extracted, respectively. CR and AVE values were calculated using the following equations [1]:

$$CR = \frac{(\sum \lambda)^2}{(\sum \lambda)^2 + \sum e} \tag{3}$$

$$AVE = \frac{\sum \lambda^2}{\sum \lambda^2 + \sum e} \tag{4}$$

Where λ is standardised factor loading and e is the standardised error.

Table 3 summarises the reliability of each factor. Both factors have a Cronbach alpha greater than 0.70, which shows measurement scales consisting of a set of homogeneous items to measure the meaning of the factor. Composite reliability for each factor in the CFA model was above 0.60, demonstrating that these factors had adequate internal consistency and were sufficient in their representation of the construct. Both factors also had AVE values greater than 0.50, indicating that more variance was captured by the variables within each factor and shared more variance in the factor than with the other factor. This also means that the construct relative to the amount of variance was due to measurement error.

Table 3. Reliability tests of the ACI construct

Factor	α	CR	AVE
Innovative culture	0.83	0.84	0.58
Performance-oriented culture	0.75	0.80	0.52

The respecified two-factor model had the acceptable model fit indices suggesting that unidimensionality could be supported. To measure scale reliability, the study utilised 'Cronbach's alpha' which provided an indication of how consistent the responses were across items within the scale. The recalculated Cronbach's alpha value of 0.849 was indicative of the reliability of the scale.

5 Conclusion

The paper presented here aimed to increase both the understanding of culture for innovation in the Australian Public Service and to improve the construct measurement scale. Perspectives on culture for innovation were identified by exploratory factor analysis. Two factors were extracted from the 11 variables which were selected and grouped according to culture type. Thus, 11 variables with factor loadings ranging from 0.620 to 0.806 were retained. The two factors extracted as characterising ambidextrous culture for innovation (ACI) were comprised of innovative culture and performance-oriented culture.

To assess the reliability and validity of the ambidextrous culture for innovation construct, a confirmatory factor analysis was conducted using the maximum likelihood estimation method. Initially, the model did not present an acceptable level of fit because variables A3, A4, A7, and A9 were associated with relatively high modification indices. Thus, it was prudent to respecify the model with the elimination of these four variables. After the model re-specification process, the CFA model fit very well with the collected data and the relationships between the observed variables and latent variables were significant. This study also provides a reliable measurement scale to measure culture for innovation in the Australian Public Service.

The two-factor model was applied both descriptively and diagnostically. It presented a practical way to measure an ambidextrous culture for innovation, and could initially be used to establish a baseline level of ambidextrous culture for innovation. The empirical evidence shows ambidextrous culture for innovation is reflected by the combination of innovative culture and performance-oriented culture. These findings highlight the importance of embedding both practices into the fabric of a government's culture. This could be achieved by senior managers promoting an innovative culture by allowing errors, encouraging employees to put forward their ideas, and cultivating a learning environment. Leaders should also facilitate a performance-oriented culture by supervising goal attainment, establishing standardised process, and adhering to plans. If successfully adopted, an ambidextrous culture for innovation will increase productivity, and may eventually result in a high performance organisation (HPO).

The validity and reliability of the construct were confirmed in the EFA and CFA, however it was still necessary to investigate for a common method bias. By taking into account the recommendations of Podsakoff et al. [19], Harman's single-factor test was conducted to assess the common method bias in this single setting design. As the results indicated no single factor underlying the data, the common method bias did not seem to exist.

References

1. Hair, J.F., et al.: Multivariate data analysis: A global perspective. Basım Pearson Education (2010)
2. Gerbing, D.W., Anderson, J.C.: An updated paradigm for scale development incorporating unidimensionality and its assessment. J. Marketing Res. **25**, 186–192 (1988)
3. Kline, R.B.: Principles and Practice of Structural Equation Modeling. Guilford Publications, New York (2015)
4. Zhou, K.Z., et al.: Developing strategic orientation in China: antecedents and consequences of market and innovation orientations. J. Bus. Res. **58**(8), 1049–1058 (2005)
5. Lægreid, P., Roness, P.G., Verhoest, K.: Explaining the innovative culture and activities of state agencies. Organ. Stud. **32**(10), 1321–1347 (2011)
6. Vartanian, T.P.: Secondary Data Analysis. Oxford University Press, New York (2010)

7. Brown, R.B., Saunders, M.P.: Dealing with Statistics: What you Need to Know. McGraw-Hill Education, UK (2007)
8. Frankfort-Nachmias, C., Nachmias, D.: Research Methods in the Social Sciences, 7th edn. Worth, New York (2008)
9. Torugsa, N., Arundel, A.: Complexity of Innovation in the public sector: a workgroup-level analysis of related factors and outcomes. Public Manag. Rev. **18**(3), 392–416 (2016)
10. Torugsa, N.A., Arundel, A.: The nature and incidence of workgroup innovation in the Australian public sector: evidence from the Australian 2011 state of the service survey. Australian J. Public Adm. **75**(2), 202–221 (2015)
11. Tabachnick, B.G., Fidell, L.S.: Using Multivariate Statistics. Allyn & Bacon, Needham Height, MA (2007)
12. Russell, D.W.: In search of underlying dimensions: the use (and abuse) of factor analysis in Personality and Social Psychology Bulletin. Pers. Soc. Psychol. Bull. **28**(12), 1629–1646 (2002)
13. Pallant, J: SPSS Survival Manual: A Step by Step Guide to Data Analysis Using IBM SPSS. McGraw-Hill, Maidenhead (2013)
14. Field, A.: Discovering Statistics Using IBM SPSS Statistics, 4th edn. SAGE Publications, London (2013)
15. O'connor, B.P.: SPSS and SAS programs for determining the number of components using parallel analysis and Velicer's MAP test. Behav. Res. Methods Instrum. Comput. **32**(3), 396–402 (2000)
16. Gallagher, D., Ting, L., Palmer, A.: A journey into the unknown; taking the fear out of structural equation modeling with AMOS for the first-time user. Mark. Rev. **8**(3), 255–275 (2008)
17. Byrne, B.M.: Structural Equation Modeling with AMOS: Basic Concepts, Applications, and Programming. Routledge, New York (2013)
18. Bagozzi, R.P., Yi, Y.: On the evaluation of structural equation models. J. Acad. Mark. Sci. **16**(1), 74–94 (1988)
19. Podsakoff, P.M., et al.: Common method biases in behavioral research: a critical review of the literature and recommended remedies. J. Appl. Psychol. **88**(5), 879 (2003)

Towards a Metaquery Language
for Mining the Web of Data

Francesca A. Lisi[(✉)]

Dipartimento di Informatica, Università degli Studi di Bari "Aldo Moro",
Bari, Italy
francesca.lisi@uniba.it

Abstract. This paper argues that Data Analytics in the Web of Data
asks for a metaquery language. One such language, based on second-
order Description Logics, is sketched and illustrated with reference to a
use case in the context of interest.

1 Introduction

The emerging *Web of Data* builds upon the WWW infrastructure to represent and interrelate data (aka *Linked Data*), with the aim of transforming the
Web from a distributed file system into a distributed database system. The
foundational standards of the Web of Data include the Uniform Resource Identifier (URI) and the Resource Description Framework (RDF)[1]. URIs are used
to identify resources and RDF is used to relate resources. RDF can be viewed
simply as a *data model* according to which data is represented in the form of
triples ⟨*subject predicate object*⟩. The resulting collection of triples is a directed,
labeled graph which can be accessed by posing SPARQL[2] queries. The link
between RDF and Description Logics (DLs) [1] allows several entailment regimes
for query answering in SPARQL.

The Web of Data will soon become a major Big Data source which appears
challenging from the viewpoint of Data Analytics. In such an open and distributed environment, Data Analytics could take advantage of some useful meta-
information about the data to be analyzed. The notion of a *metaquery* was proposed in the 90s as a template that describes the type of pattern to be discovered
in relational databases [2]. A distinguishing feature of metaquerying problems is
the use of a second-order logic language. In this paper, we argue that a similar
approach should be followed for the case of mining the Web of Data. To this purpose we briefly report on a metaquery language based on second-order DLs but
implementable with standard technologies underlying the Web of Data, notably
SPARQL.

The paper is structured as follows. Section 2 provides a motivation for this
investigation by describing a possible use case for metaquerying in the Web of
Data context. Section 3 sketches our metaquery language. Section 4 concludes
the paper with final remarks.

---

[1] https://www.w3.org/RDF/.
[2] https://www.w3.org/TR/rdf-sparql-query/.

ⓒ Springer International Publishing AG 2017
A. Calì et al. (Eds.): BICOD 2017, LNCS 10365, pp. 90–93, 2017.
DOI: 10.1007/978-3-319-60795-5_8

2   A Motivating Use Case

Most sources in the current Web of Data are so-called *knowledge graphs* (KGs), *i.e.* huge collections of RDF triples. A typical case of a large KG is *DBPedia*,[3] which, essentially, makes the content of Wikipedia available in RDF and incorporates links to other datasets on the Web, *e.g.*, to *Geonames*[4]. RDF triples can be straightforwardly represented by means of unary and binary first-order logic (FOL) predicates. More precisely, the unary predicates are the objects of the RDF *type* predicate, while the binary ones correspond to all other RDF predicates, *e.g.*, ⟨*alice type researcher*⟩ and ⟨*bob isMarriedTo alice*⟩ from the KG in Fig. 1 refer to *researcher*(*alice*) and *isMarriedTo*(*bob*, *alice*) respectively.

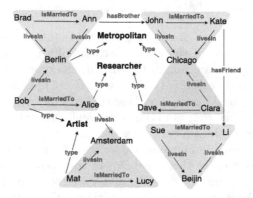

Fig. 1. Fragment of a knowledge graph (taken from [3]).

Since KGs are automatically constructed by applying information extraction techniques, they are inherently *incomplete*. Therefore, the task of *completion* (also known as *link prediction*) is of crucial importance for the curation of KGs. To this aim, data mining algorithms can be exploited to automatically build rules able to make predictions on missing links. For example, the following rule

$$r1 : isMarriedTo(x, y), livesIn(x, z) \Rightarrow livesIn(y, z) \tag{1}$$

can be mined from the KG in Fig. 1 and applied to derive new facts such as *livesIn*(*alice*, *berlin*), *livesIn*(*dave*, *chicago*) and *livesIn*(*lucy*, *amsterdam*) to be used for completing the graph.

Many results from the data mining literature, including the works on metaquery languages, can be extended to the Web of Data. However, this is not straightforward. Indeed, due to their inherent incompleteness, KGs are naturally treated under the Open World Assumption (OWA) as opposed to databases for which the Closed World Assumption (CWA) holds. This motivates our proposal of a metaquery language based on second-order DLs.

---

[3] http://wiki.dbpedia.org/.
[4] http://www.geonames.org/.

3   Metaqueries with Second-Order DLs

Let $\mathcal{DL}$ be any DL with syntax $(\mathsf{N}_\mathcal{C}, \mathsf{N}_\mathcal{R}, \mathsf{N}_\mathcal{O})$ where $\mathsf{N}_\mathcal{C}$, $\mathsf{N}_\mathcal{R}$, and $\mathsf{N}_\mathcal{O}$ are the alphabet of concept names, role names and individual names, respectively.

First, we extend $\mathcal{DL}$ so that we can go far beyond the standard way of querying DL knowledge bases (KBs), *i.e.*, instance retrieval. For this purpose, let $\mathsf{V}_\mathcal{O}$ be a countably infinite set of *individual variables* disjoint from $\mathsf{N}_\mathcal{C}$, $\mathsf{N}_\mathcal{R}$, and $\mathsf{N}_\mathcal{O}$. A *term* t is an element from $\mathsf{V}_\mathcal{O} \cup \mathsf{N}_\mathcal{O}$. Let C be a concept, r a role, and t, t' terms. An *atom* is an expression $C(t)$, $r(t, t')$, or $t \approx t'$ and we refer to these three different types of atoms as *concept atoms*, *role atoms*, and *equality atoms* respectively. A *conjunctive query* (CQ) is a non-empty set of atoms such as

$$\{isMarriedTo(x, y), livesIn(x, z), livesIn(y, z)\} \tag{2}$$

where we assume that x is a *distinguished* variable (also called answer or free variable) and y, z are non-distinguished (existentially quantified) variables. If all variables in the query are non-distinguished, the query answer is just true or false and the query is called a Boolean CQ. If a CQ contains distinguished variables, the answers to the query are those tuples of individual names for which the KB entails the query that is obtained by replacing the free variables with the individual names in the answer tuple.

Since we are interested in metaqueries, we need to introduce two further sets of variables (of second-order this time): $\mathsf{V}_\mathcal{C}$ of so-called *concept variables*, *i.e.* variables that can be quantified over $\mathsf{N}_\mathcal{C}$, and $\mathsf{V}_\mathcal{R}$ of so-called *role variables*, *i.e.* variables that can be quantified over $\mathsf{N}_\mathcal{R}$. Let then $\mathcal{DL}_\chi$ be the second-order DL language obtained by extending $\mathcal{DL}$ with $\mathsf{V}_\mathcal{C}$ and $\mathsf{V}_\mathcal{R}$. Intuitively, a metaquery is a second-order CQ. So, the notions of term and atom introduced above should be extended to the second-order case. However, for the purpose of this work, we can restrict our language to particular second-order CQs, *e.g.* involving role variables and individual variables such as in

$$\{P(x, y), Q(x, z), Q(y, z)\} \tag{3}$$

which looks for properties (Q) shared by two individuals that are related (P). Note that (2) is a possible instantiation of (3) and can be the starting point for the generation of several rules such as (1). Of course, whether the pattern represented by (2) is interesting or not is a matter for evaluation by means of ad-hoc measures to be defined.

As for the semantics of $\mathcal{DL}_\chi$, we propose to follow the Henkin style. A nice feature of the *Henkin semantics* [4], as opposed to the Standard Semantics, is that the expressive power of the language actually remains first-order. This paves the way for the use of first-order solvers in spite of the second-order syntax. Also, this is a shared feature with RDF(S). Last, but not least, it makes possible an implementation with SPARQL.

4   Conclusions

The issue of extending DLs with higher-order constructs for metamodeling purposes has hardly been addressed. However, since the importance of metamodeling (of which metaquerying is a special case) in several applications has been recognized, there is an increasing interest in the topic [5–7]. In particular, De Giacomo *et al.* [7] augment a DL with variables that may be interpreted - in a Henkin semantics - as individuals, concepts, and roles at the same time, obtaining a new logic $Hi(\mathcal{DL})$. Colucci *et al.* [8] introduce second-order features in DLs under Henkin semantics for modeling several forms of non-standard reasoning. Lisi [9] extends [8] to some variants of concept learning, thus being the first to propose higher-order DLs as a means for metamodeling in Data Analytics.

In this paper we have sketched a metaquery language for mining the Web of Data. It is based on second-order DLs but implementable with SPARQL. However, several aspects need to be clarified before an implementation. The most prominent ones are the instantiation mechanisms and the evaluation measures.

References

1. Baader, F., Calvanese, D., McGuinness, D., Nardi, D., Patel-Schneider, P. (eds.): The Description Logic Handbook: Theory, Implementation and Applications, 2nd ed. Cambridge University Press (2007)
2. Shen, W., Ong, K., Mitbander, B.G., Zaniolo, C.: Metaqueries for data mining. In: Advances in Knowledge Discovery and Data Mining, pp. 375–398 (1996)
3. Dang Tran, H., Stepanova, D., Gad-Elrab, M., Lisi, F.A., Weikum, G.: Towards nonmonotonic relational learning from knowledge graphs. In: Cussens, J., Russo, A. (eds.) Proceedings of the 26th International Conference on Inductive Logic Programming (ILP 2016). LNCS, vol. 10326, Springer (2017). under publication
4. Henkin, L.: Completeness in the theory of types. J. Symbolic Logic **15**(2), 81–91 (1950)
5. Pan, J.Z., Horrocks, I.: OWL FA: a metamodeling extension of OWL DL. In: Carr, L., De Roure, D., Iyengar, A., Goble, C.A., Dahlin, M. (eds.) Proceedings of the 15th International Conference on World Wide Web, WWW 2006, Edinburgh, Scotland, UK, 23–26 May 2006, pp. 1065–1066. ACM (2006)
6. Motik, B.: On the properties of metamodeling in OWL. J. Logic Comput. **17**(4), 617–637 (2007)
7. De Giacomo, G., Lenzerini, M., Rosati, R.: Higher-order description logics for domain metamodeling. In: Burgard, W., Roth, D. (eds.) Proceedings of the 25th AAAI Conference on Artificial Intelligence, AAAI 2011, San Francisco, California, USA, 7–11 August 2011 (2011)
8. Colucci, S., Di Noia, T., Di Sciascio, E., Donini, F.M., Ragone, A.: A unified framework for non-standard reasoning services in description logics. In: Coelho, H., Studer, R., Wooldridge, M. (eds.) ECAI 2010–19th European Conference on Artificial Intelligence, Lisbon, Portugal, 16–20 August 2010, Proceedings. Frontiers in Artificial Intelligence and Applications, vol. 215, pp. 479–484. IOS Press (2010)
9. Lisi, F.A.: A declarative modeling language for concept learning in description logics. In: Riguzzi, F., Železný, F. (eds.) ILP 2012. LNCS (LNAI), vol. 7842, pp. 151–165. Springer, Heidelberg (2013). doi:10.1007/978-3-642-38812-5_11

Enabling Deep Analytics in Stream Processing Systems

Milos Nikolic[1]([✉]), Badrish Chandramouli[2], and Jonathan Goldstein[2]

[1] University of Oxford, Oxford, UK
milos.nikolic@cs.ox.ac.uk
[2] Microsoft Research, Redmond, USA
badrishc@microsoft.com, jongold@microsoft.com

Abstract. Real-time applications often analyze data coming from sensor networks using relational and domain-specific operations such as signal processing and machine learning algorithms. To support such increasingly important scenarios, many data management systems integrate with numerical frameworks like R. Such solutions, however, incur significant performance penalties as relational engines and numerical tools operate on fundamentally different data models with expensive inter-communication mechanisms. In addition, none of these solutions supports efficient real-time and incremental analysis. In this work, we advocate a deep integration of domain-specific operations into general-purpose query processors with the goal of providing unified query and data models for both online and offline processing. Our proof-of-concept system tightly integrates relational and digital signal processing operations and achieves orders of magnitude better performance than existing loosely-coupled data management systems.

1 Introduction

Increasingly many applications analyze data that comes from large networks of sensor devices, commonly known as the Internet of Things (IoT). Typical IoT applications combine relational and domain-specific operations in processing sensor signals: for example, the former group signals by different sources or join signals with historical data; the latter use Fast Fourier Transform (FFT) to do spectral analysis, digital filters to recover noisy signals, or online machine learning to classify signal values. Reconciling these two seemingly disparate worlds, especially for real-time analysis, is challenging.

The database community has recognized the need for a tighter integration of data management systems and domain-specific algorithms. Numerical computing environments such as MATLAB and R provide efficient domain-specific algorithms but remain unsuitable for general-purpose data processing. In order to enable specialized routines in complex workflows, many data management systems nowadays integrate with numerical frameworks, R in particular: Spark [4] and SciDB [1] provide R packages; SQL Server, MonetDB, and StreamBase support queries that can invoke R code.

© Springer International Publishing AG 2017
A. Calì et al. (Eds.): BICOD 2017, LNCS 10365, pp. 94–98, 2017.
DOI: 10.1007/978-3-319-60795-5_9

However, the existing integration mechanisms between database systems and numerical frameworks are suboptimal performance-wise as they treat both sides as independent systems. Such *loose system coupling* comes with significant processing overheads – for instance, executing R programs requires exporting data from the database, converting into R format, running R scripts, converting back into a relational format, and importing into the database. Sending data back and forth between the systems often dominates the execution time and increases latency, which makes this approach particularly unsuitable for real-time processing.

2 Domain-Specific Computation in Stream Processing Engines

We advocate a *deep integration* of domain-specific operations with general-purpose stream processors. This approach aims to bring specialized operations closer to data and eliminate the need for expensive communication with external numerical tools. Enabling in-situ domain-specific computation in stream engines empowers users to express end-to-end workflows more succinctly, inside one system and using one language.

This tight integration poses several requirements and challenges:

1. *Query and data model reconciliation.* General-purpose stream processing engines and numerical tools use different query and data models. The former support relational and streaming queries over relational or tempo-relational data; the latter support domain-specific, mostly offline, computations on arrays. The key challenge is how to seamlessly unify these disparate models instead of simulating them on top of each other, and yet provide data practitioners and domain experts with familiar abstractions.

2. *Extensibility.* A query processor supporting deep analytics should allow domain experts to implement custom operators in a way that feels natural to them – by writing algorithms against arrays without worrying about the format of the underlying data. Exposing arrays to operator writers would enable easy integration of existing highly optimized algorithms, for instance, implementations using SIMD instructions.

3. *Online and incremental computation.* Stream processors loosely coupled with MATLAB or R cannot incrementalize domain-specific tasks that operate over hopping (overlapping) windows of data. The stateless nature of routines in MATLAB and R forces stream engines to redundantly recompute over overlapping subsets of data. On the other hand, deep integration admits more efficient data management (e.g., via fixed-size circular arrays) and incremental computation in user-defined stateful operators.

4. *Performance.* High performance is always a critical requirement for analytics. The deep integration approach can bring more expressiveness to the query language but carries a risk of completely giving up on performance. In order to be welcomed by data scientists and domain experts, a deeply integrated

system should preserve the performance of existing relational operators while being competitive with best-of-breed systems for domain-specific tasks. For mixed workflows, a tightly-coupled system should exhibit better performance than existing loosely-coupled alternatives.

3 Proof of Concept: Signal Processing over Data Streams

Our recent work presents a system that deeply integrates relational and digital signal processing (DSP) while satisfying the above requirements [3]. The system, called TRILLDSP, is built on top of Trill [2], a high-performance incremental analytics engine that supports processing of streaming and relational queries.

Trill uses a tempo-relational data model to uniformly represent offline and online datasets as stream data. Each stream event is associated with a time window that denotes its period of validity. Such stream events form snapshots of valid data versions across time. The user query is executed against these snapshots in an incremental manner. Trill's query language provides standard relational operators (e.g., selection, projection, join, etc.) with temporal interpretation. Each operator is a function from stream to stream, which allows for elegant functional composition of queries.

TRILLDSP provides a unified query language for processing tempo-relational and signal data. We next show a query that combines these seemingly disparate worlds.

Example 1. An IoT application receives a stream of temperature readings from different sensors in time order. Each reading has the same format `<SensorId,Time,Value>`. The application runs the same query on every signal coming from a different source.

```
var q = stream.Map(s => s.Select(e => e.Value), e => e.SensorId)
           .Reduce(s => s.Window(512, 256,
               w => w.FFT().Select(a => f(a)).InverseFFT(), a => a.Sum())));
```

The query combines relational and DSP operators to express several processing phases: (1) *Grouping:* `Map` specifies a sub-query (projection with `Select`) to be performed in parallel on the stream and a grouping key (`SensorId`) to be used for shuffling the result streams; `Reduce` specifies the query to be executed per each group; (2) *Windowing:* Most DSP algorithms operate over windows of data defined by window size and hop size. The `Window` operator allows users to express such algorithms as a series of transformations of fixed-size arrays (`w`); (3) *Spectral Analysis:* The processing pipeline starts with a Fast Fourier Transform (FFT) that computes the frequency representation of a 512-sample window at each hopping point. A user-defined function f modifies the computed spectrum (e.g., retains top-k spectrum values and zeros out others) before invoking an inverse FFT; (4) *Unwindowing:* To restore the signal form, the final phase projects output arrays back to the time axis and sums up the overlapping values.

This query captures the fundamental technique of windowing in DSP and serves as a blueprint for a large class of IoT workflows. The declarative query model of TRILLDSP offers high-level operators capable of expressing such deep analytics succinctly.   □

We design and implement the DSP functionality in TRILLDSP as a layer running on top of the unmodified Trill relational engine. The layer enriches the query model with a "walled garden" that provides signal abstractions and clear transitions between stream and signal operations. TRILLDSP makes no changes in the underlying tempo-relation data model (i.e., has no arrays as first-class citizens). Instead, it exposes arrays only through designated DSP operators and abstracts away the complexity of on-the-fly transformations between the relational and array models. In that way, the existing relational API and the performance of non-DSP queries remain unaffected.

For DSP experts, TRILLDSP provides an extensible framework for defining new user-defined window operations (e.g., variants of FFT, windowing functions, correlation functions, element-wise product, etc.), which can be seamlessly interleaved with relational operators. The framework internally exposes array abstractions to ease the integration of new black-box DSP operators. It also allows users to implement incremental versions of operators to be used in computation with hopping windows.

The unified query model of TRILLDSP supports both online and offline analysis. As offline datasets are streams consisting of events with an infinite lifetime, each tempo-relational operator is transferable between real-time and offline by construction. Thus, users can build queries from offline data and then put them unmodified in production.

Deep integration of signal and relational processing is key to achieving high performance. In pure DSP tasks, TRILLDSP is competitive or even faster than state-of-the-art numerical frameworks like MATLAB and R. This comes at no surprise as this framework exposes array abstractions to operator writers for easy adoption of highly-optimized black-box implementations of DSP operations (e.g., those exploiting SIMD instructions). TRILLDSP shows its full potential when processing data coming from a large number of sensor devices using the query logic that combines relational and signal operation. All signal operations natively support grouped processing, that is, one operator can simultaneously process multiple groups by maintaining the state of each group. Coupling these operators with Trill's streaming temporal MapReduce operator enables efficient grouped signal processing. TRILLDSP's in-situ execution model achieves up to two orders of magnitude better performance than modern relational and array data management systems with loose R integration, such as SPARKR [4] and SCIDB-R [1].

4   Future Work

This proof-of-concept system [3] demonstrates that one can extend general-purpose query processors with domain-specific functionality and get the best of both worlds: a more powerful, declarative query language and high performance of both general-purpose and domain-specific operations. This result opens up the question whether the "one-size-doesn't-fit-all" paradigm from database systems also holds for streaming systems because of their dynamic nature that allows them to adapt data on-the-fly to meet different processing requirements.

The "walled garden" approach to enabling deep analytics over data streams can also be applied in domains other than signal processing, such as machine learning, scientific computing, computer vision, physical modeling, computational finance, etc. The window-based query template shown in the above example can support arbitrary computations over arrays: for instance, applications analyzing accelerometer data from wearable devices can build pipelines of machine learning operators to classify sequences of signal samples into classes of physical activities (e.g., walking, running, sitting, standing, etc.); algorithmic trading applications can implement trading strategies using pipeline operators over windows of stock prices, which can be possibly correlated with historical data. In general, enhancing relational engines with powerful capabilities from various domains enables data processing to be placed closer to data sources (i.e., towards network edges), which soon will be essential in order to cope with growing amounts of sensor data and ensure scalability of IoT applications.

References

1. Brown, P.G.: Overview of SciDB: large scale array storage, processing and analysis. In: SIGMOD, pp. 963–968 (2010)
2. Chandramouli, B., Goldstein, J., Barnett, M., DeLine, R., Platt, J.C., Terwilliger, J.F., Wernsing, J.: Trill: a high-performance incremental query processor for diverse analytics. PVLDB 8(4), 401–412 (2014)
3. Nikolic, M., Chandramouli, B., Goldstein, J.: Enabling signal processing over data streams. In: SIGMOD, pp. 95–108 (2017)
4. Zaharia, M., Chowdhury, M., Das, T., Dave, A., Ma, J., McCauly, M., Franklin, M.J., Shenker, S., Stoica, I.: Resilient distributed datasets: a fault-tolerant abstraction for in-memory cluster computing. In: NSDI, pp. 15–28 (2012)

Graph Data Querying and Analysis

Link Prediction Using Top-k Shortest Distances

Andrei Lebedev, JooYoung Lee, Victor Rivera$^{(\boxtimes)}$, and Manuel Mazzara

Innnopolis University, Innopolis, Russia
{a.lebedev,j.lee,v.rivera,m.mazzara}@innopolis.ru

Abstract. Top-k shortest path routing problem is an extension of finding the shortest path in a given network. Shortest path is one of the most essential measures as it reveals the relations between two nodes in a network. However, in many real world networks, whose diameters are small, top-k shortest path is more interesting as it contains more information about the network topology. In this paper, we apply an efficient top-k shortest distance routing algorithm to the link prediction problem and test its efficacy. We compare the results with other base line and state-of-the-art methods as well as with the shortest path. Our results show that using top-k distances as a similarity measure outperforms classical similarity measures such as Jaccard and Adamic/Adar.

Keywords: Graph databases · Shortest paths · Link prediction · Graph matching · Similarity

1 Introduction

In a connected world, emphasis is on relationships more than isolated pieces of information. Relational databases may compute relationships at query time [13,14], but this would result in a computationally expensive solution. Graph databases [3] store connections as first class citizens, allowing access to persistent connections in almost constant-time [15]. One of the fundamental topological features in the context of graph theory and graph databases, with implications in Artificial Intelligence and Web communities [5], is the computation of the shortest-path distance between vertices [10].

Many efficient methods for finding shortest paths have been proposed. On the other hand, top-k distance query handling methods are not well developed and spread. They have many advantages over traditional shortest path as they reveal much more information over a simple shortest path [1]. To extract top-k distances from graph databases, an efficient indexing algorithm is needed, such as the pruned landmark labeling scheme, presented in [2]. We utilize this algorithm to obtain the distances and then develop a similarity metric based on them to predict links on graphs.

Table 1 summarizes the connections presented in the graph depicted in Fig. 1. Based on top-1 distance, we can observe that the similarity (the shortest distance) between black nodes (i.e. $\{a, b\}$, $\{b, c\}$) is the same in the graph. However,

© Springer International Publishing AG 2017
A. Calì et al. (Eds.): BICOD 2017, LNCS 10365, pp. 101–105, 2017.
DOI: 10.1007/978-3-319-60795-5_10

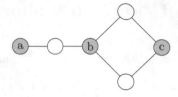

Fig. 1. An example of connections between pairs of vertices.

it is clear that nodes $\{b, c\}$ are connected more tightly through a greater number of shortest paths.

The framework described in this paper is based on previous work which is called Pruned Landmark Labeling [2]. We evaluate and test the proposed algorithm extensively to prove the efficacy of the algorithm. Then we apply the algorithm to the link prediction problem to compare with existing solutions.

Table 1. Distances and top-k distances between pairs of black vertices in Fig. 1.

Vertex pair	(Top-1) distance	Top-k distance
(a,b)	2	[2, 4, 6 ...]
(b,c)	2	[2, 2, 4 ...]

2 Related Work

One of the attempts to find k shortest paths is presented in [4] which achieves $\mathcal{O}(m + n \log n + k)$ complexity. [4] also covers programming problems, including the knapsack problem, sequence alignment, maximum inscribed polygons, and genealogical relationship discovery. We adopt the algorithm presented in [1] to discover k shortest paths since it achieves six orders of magnitude faster computation given very large graphs with millions of vertices and tens of millions of edges.

Link prediction in social networks is a well known problem and extensively studied. Links are predicted using either semantic or topological information of a given network. The main idea in the link prediction problem is to measure the similarity between two vertices which are not yet linked to each other. If the measured similarity is high enough then the future link is predicted. The work in [11] represents an attempt to infer which new interactions among members of a social network are likely to occur in the near future. The authors develop approaches to link prediction based on measures for analyzing the "proximity" of nodes in a network. Link prediction has concrete applications in trust-based social networks, where interactions that are most-likely to occur in future can influence trust metrics [12].

3   Prediction Method

Given a graph $G = (V, E)$, where V is the set of vertices and E is the set of edges. let m and n be $|E|$ and $|V|$ respectively. We assume that vertices are represented by unique integers to enable the comparison of two vertices. Furthermore, let us denote by $\mathcal{P}_{ab}$ the set of paths from a to b, where $a \in V$ and $b \in V$, and $d_{i^{th}}(s, t)$ as the i-th shortest path in $\mathcal{P}_{st}$.

In the following, we introduce the structure of the querying and indexing algorithm from [1].

- *Distance label $L(v)$*: a set of triplets (u, δ, c) of a vertex, a path length and a number of paths with length δ.
- *Loop label $C(v)$*: a sequence of k integers $(\delta_1, \delta_2, ..., \delta_k)$.
- *Index $I = (L, C)$*: L and C sets of distance labels and loop labels.
- *Ordering*: vertices in order of decreasing degrees.
- *Number of paths $c_{w,\delta'}$*: number of paths in the $L(v)$ between vertex v and w, of length not exceeding δ' using loop label $C(v)$.
- Query(I, s, t): smallest k elements in the $\Delta(I, s, t)$.

Then we can compute the multiset as follows.

$$\Delta(I, s, t) = \{\delta_{sv} + \delta_{vv} + \delta_{vt} \mid (v, \delta_{sv}) \in L(s), \delta_{vv} \in C(v), (v, \delta_{vt}) \in L(t)\}$$

Referring to the original work by [1], we have measured the performance in two ways: index construction speed and the final index size. Our implementation which considers unweighted and undirected graphs has achieved a reduction in index size compared to [1].

Proposed Method: First, we implemented the algorithm presented in [1] to compute top-k shortest paths between two vertices. Then we use the sum of top-k shortest paths as the similarity measure to predict future links. This naive approach shows better results compared to other commonly used link prediction methods.

$$S_k = \Sigma_{i=0}^{k-1} KSP(s, t, k)[i] \tag{1}$$

Equation 1 shows the similarity measure based on top-k distances where $KSP(s, t, k)$ is the list of top-k shortest paths between vertices s and t.

4   Experimental Results

Setup: In these experiments, all networks were treated as undirected unweighted graphs without self-loops and multiple edges. For testing purposes, we randomly sample 60% of edges for prediction and 40% for evaluation. The sampling, prediction and evaluation tasks were performed 10 times on each dataset. We use AUROC (Area Under the ROC curve) as an evaluation metric. We used five different datasets from [1].

Table 2. Performance evaluation of predictions on 5 datasets. Statistics of each graph are described in [1].

	Top1	Top4	Top16	Top64	CN	Jaccard	Adamic	Preferential
Facebook-1	0.878481	**0.909458**	0.899959	0.886813	0.834086	0.833845	0.799192	0.693485
Last.fm	0.863925	**0.88316**	0.881586	0.87527	0.736351	0.733341	0.721282	0.782929
GrQc	0.853479	**0.853527**	0.851746	0.84664	0.784875	0.784865	0.726173	0.720467
HepTh	0.826561	**0.82677**	0.824997	0.819686	0.730995	0.730984	0.671017	0.690331
CondMat	**0.911328**	0.911099	0.90812	0.901543	0.815252	0.815241	0.75657	0.716568

Results: The performance of our proposed method is summarized in Table 2. The best performance for each dataset is emphasized. As we can see, Top-4 shortest distance consistently performs better than the others except *CondMat* in which the difference is negligible. Intuitively, one might think bigger k should be a better predictor but our results suggest that small k is enough. Even with a small k value such as 4, we can predict future links more effectively. From the experiments, we can conclude that top-k distances capture the structural similarity between vertices better than commonly used measures, namely, *Common Neighbors (CN), Jaccard, Adamic/Adar* and *Preferential Attachment*.

5   Conclusions and Future Work

In this paper, we defined a new similarity metric between two users of social networks based on top-k shortest distances. We also found out through experiments that our metric outperforms other common metrics and also a small k suffices to accurately predict future links. Since top-k distances capture important topological properties between vertices, we plan to apply the metric in gene regulation networks to discover unknown relationships among genes which are difficult to infer using other methods. Also it would be interesting to utilize top-k distances to observe the spread of reputation (or, in general, information) in social networks since many studies suggest that reputation is transferable [7–9].

Furthermore, graph databases are a growing technology and, in some cases, shortest path implementation is already at their core (for example, Neo4j [6]). It is natural therefore to investigate the results in the context of this development, in order to identify possible improvements in performance gaps.

References

1. Akiba, T., Hayashi, T., Nori, N., Iwata, Y., Yoshida, Y.: Efficient top-k shortest-path distance queries on large networks by pruned landmark labeling. In: AAAI 2015, pp. 2–8 (2015)
2. Akiba, T., Iwata, Y., Yoshida, Y.: Fast exact shortest-path distance queries on large networks by pruned landmark labeling. In: SIGMOD 2013, pp. 349–360 (2013)
3. Angles, R., Gutierrez, C.: Survey of graph database models. ACM Comput. Surv. **40**(1), 1:1–1:39 (2008)

4. Eppstein, D.: Finding the k shortest paths. SIAM J. Comput. **28**(2), 652–673 (1999). doi:10.1137/S0097539795290477
5. Goldberg, A.V., Harrelson, C.: Computing the shortest path: a search meets graph theory. In: SODA 2005, pp. 156–165 (2005)
6. Lal, M.: Neo4J Graph Data Modeling. Packt Publishing, Birmingham (2015)
7. Lee, J.Y., Oh, J.C.: A model for recursive propagations of reputations in social networks. In: Proceedings of the 2013 IEEE/ACM International Conference on Advances in Social Networks Analysis and Mining, pp. 666–670. ACM (2013)
8. Lee, J., Duan, Y., Oh, J.C., Du, W., Blair, H., Wang, L., Jin, X.: Automatic reputation computation through document analysis: a social network approach. In: International Conference on Advances in Social Networks Analysis and Mining (ASONAM), pp. 559–560. IEEE (2011)
9. Lee, J., Oh, J.C.: Convergence of true cooperations in bayesian reputation game. In: IEEE 13th International Conference on Trust, Security and Privacy in Computing and Communications (TrustCom), pp. 487–494. IEEE (2014)
10. Lee, J., Oh, J.C.: Estimating the degrees of neighboring nodes in online social networks. In: Dam, H.K., Pitt, J., Xu, Y., Governatori, G., Ito, T. (eds.) PRIMA 2014. LNCS, vol. 8861, pp. 42–56. Springer, Cham (2014). doi:10.1007/978-3-319-13191-7_4
11. Liben-Nowell, D., Kleinberg, J.: The link-prediction problem for social networks. J. Am. Soc. Inform. Sci. Technol. **58**(7), 1019–1031 (2007)
12. Mazzara, M., Biselli, L., Greco, P.P., Dragoni, N., Marraffa, A., Qamar, N., de Nicola, S.: Social networks and collective intelligence: a return to the agora. IGI Global (2013)
13. Qu, Q., Chen, C., Jensen, C.S., Skovsgaard, A.: Space-time aware behavioral topic modeling for microblog posts. IEEE Data Eng. Bull. **38**(2), 58–67 (2015)
14. Qu, Q., Liu, S., Yang, B., Jensen, C.S.: Integrating non-spatial preferences into spatial location queries. In: SSDBM 2014, pp. 8:1–8:12 (2014)
15. Robinson, I., Webber, J., Eifrem, E.: Graph Databases. O'Reilly Media Inc., Sebastopol (2013)

Optimisation Techniques for Flexible Querying in SPARQL 1.1

Riccardo Frosini[✉]

Department of Computer Science and Information Systems,
Birkbeck University of London, London, UK
riccardo@dcs.bbk.ac.uk

Abstract. Flexible querying techniques can be used to enhance users' access to heterogeneous data sets, such as Linked Open Data. An extension of the SPARQL query language with flexible capabilities, called SPARQL^{AR}, was defined in [2], where an evaluation algorithm based on query rewriting was presented. In this paper, we propose two optimisation techniques for evaluating SPARQL^{AR} queries. The first is based on a caching technique, in which we pre-compute some answers in advance. The second exploits a summarisation of the data graph being queried.

1 Introduction

Linked Open Data (LOD) is a growing movement for publishing and interlinking web data using the Resource Description Framework (RDF). The amount of LOD is increasing every year[1] with multiple heterogeneous datasets being added and publicly available. Due to the size and heterogeneity of LOD, users might find it hard to query such data: they may lack full knowledge of the structure of the data, its irregularities, and the resources used within it. Moreover, schemas and resources can evolve over time, making it difficult for users to formulate queries that precisely express their information retrieval requirements. Flexible querying techniques can aid users in posing queries over such complex datasets.

An RDF dataset is a graph where the nodes and edges are resources denoted by URIs. The edges of an RDF graph are known as properties. The predominant language for querying RDF is SPARQL, whose latest version, SPARQL 1.1, supports regular path queries (also known as property paths[2]). For example, the following SPARQL 1.1 query returns every author who co-authored with "John Smith" and their dates of birth:

```
SELECT ?c ?n
WHERE { <John.Smith> <authorOf>/<writtenBy> ?c . ?c <dateOfBirth> ?n }
```

Flexible SPARQL Querying. An extension of SPARQL 1.1 with flexible capabilities has been proposed in [2], where two new operators, APPROX and RELAX,

---

[1] http://lod-cloud.net/.
[2] http://www.w3.org/TR/sparql11-property-paths/.

© Springer International Publishing AG 2017
A. Calì et al. (Eds.): BICOD 2017, LNCS 10365, pp. 106–110, 2017.
DOI: 10.1007/978-3-319-60795-5_11

are added to the language: this new language is called SPARQL^{AR}. These operators can be applied to individual conjuncts of a SPARQL 1.1 query. The conjuncts to which we do not apply APPROX or RELAX are termed the *exact part* of the query. The APPROX operator edits the property path of a query conjunct by deleting, replacing or inserting properties. For insertion, the symbol _ represents the disjunction of all properties in the dataset. The RELAX operator undertakes ontology-driven relaxation, such as replacing a property by a super-property, or a class by a super-class.

This kind of combination of query approximation and relaxation was first proposed in [3] in the context of conjunctive regular path queries (but not SPARQL). In [2], Calì et al. define the semantics of SPARQL^{AR} and investigate its complexity. The evaluation of a SPARQL^{AR} query Q over an RDF graph G and ontology K, $[[Q]]_{G,K}$, returns a set of pairs $\langle ans, cost \rangle$, where the answers are ranked according to their cost. Calì et al. propose a rewriting algorithm for evaluating a SPARQL^{AR} query Q which generates a set of pairs $\langle q, c \rangle$: q is a SPARQL 1.1 query and c is the cost associated with q. The algorithm works as follows: For each conjunct of query Q, a new query is constructed by applying one step of approximation or relaxation and each such query is assigned the cost of the approximation/relaxation. From each query generated in this way, a new set of queries is next produced by applying a second step of approximation or relaxation. This process continues for a bounded number of approximation/relaxation steps. The cost assigned to each query generated is the cost of the sequence of approximations/relaxations applied to Q derive it. For practical reasons, the number of queries generated is limited by bounding their cost up to a maximum value. For each pair $\langle q, c \rangle$ generated by the rewriting algorithm, the query q is evaluated and the cost c is assigned to its answers. Finally, all answers are returned ranked according to their cost.

In [2] the authors also present a performance study and an optimisation technique based on pre-computation. Since the rewriting algorithm generates many syntactically similar queries, the pre-computation evaluates the exact part of the query once and stores it in a *cache*. Then the approximated and relaxed conjuncts of the query are evaluated individually, and also in pairs; the answers to these are also stored in the cache. To avoid memory overflow, an upper limit is placed on the size of cache. For each query to be evaluated, the evaluator checks if parts of the query can be reused from the cache.

Two New Optimisations. Although this optimisation helps to speed up query performance, there are still problems due to the large number of queries generated by the rewriting algorithm that need to be evaluated. Moreover, many of these queries return no answers due to the edit operations made by APPROX. Hence, we propose here two optimisation techniques to further speed up the evaluation of SPARQL^{AR} queries. Our first optimisation is an improvement to the pre-computation technique of [2], where instead of caching the answers returned by single conjuncts and pairs of conjuncts, we consider also larger sets of conjuncts, so as to avoid the computation of Cartesian products.

Our second optimisation technique is based on graph summarisation. In [2] many queries generated by the rewriting algorithm return no answers. Hence, we propose the construction of a summary S of an RDF graph G such that the following property holds: if the answer of a query Q over S is empty, then the answer of Q over G is empty. Since the summary is smaller than the original graph G, our aim is to speed up query evaluation by avoiding the execution of queries that return no answers over G.

2   Query Pre-computation

In contrast to the earlier version of this optimisation mentioned above, we do not explicitly separate the query into two parts, the exact and the approximated/relaxed part, since this might lead to the calculation of the Cartesian product of sets of pre-computed answers. Instead, for each query Q generated by the rewriting algorithm, we generate a set of queries QS. Each query $Q' \in QS$ contains the exact part of query Q plus one or more non-exact triple patterns such that query Q' is connected, i.e. if Q' is split into any two sub-queries these share a variable. QS contains all the sub-queries of Q that satisfy these conditions. Each query in QS is evaluated and stored in the cache, unless the cache is full. We then construct the answers of Q by utilising the answers stored in the cache. If parts of a query $Q' \in QS$ have already been computed, then we reuse these answers to compute Q'.

We have conducted preliminary performance studies on this pre-computation optimisation technique which indicate that, by using this optimisation, we are able to reduce the time for evaluating queries that have a large number of rewritings. However, this optimisation does not help with queries that have a lower number of rewritings, but instead increases their execution time somewhat. This is due to the additional time that the pre-computation algorithm takes to compute partial queries and store their answers in the cache.

3   Graph Summarisation

Due to the nature of the APPROX operator, the rewriting algorithm might generate many queries that do not return any answer. Hence we propose an optimisation that detects and discards queries that are unsatisfiable.

The *summary* of an RDF graph G is an automaton R that is able to recognise strings that represent paths in G up to a certain length $n \geq 2$. We can show that the set of strings generated by G, $\mathcal{L}(G)$, is a subset of the set of strings generated by the summary R, $\mathcal{L}(R)$. The states in R keep track of the last k transitions that have been traversed, for all $k < n$. So the automaton will keep track only of the last $n - 1$ states even if we have traversed more than $n - 1$ states. The automaton R that recognises paths of G up to length n is constructed as follows:

1. Initially R contains only one state, S, which is both initial and final.
2. For each $p_1 p_2 \ldots p_k \in \mathcal{L}(G)$ with $k < n$, we add the new final states $S_{p_1}, \ldots,$ $S_{p_1 \ldots p_k}$ to R, and also the new transitions: $(S, p_1, S_{p_1}), (S_{p_1}, p_2, S_{p_1 p_2}), \ldots,$ $(S_{p_1 \ldots p_{k-1}}, p_k, S_{p_1 \ldots p_k})$.
3. For each $p_1 p_2 \ldots p_n \in \mathcal{L}(G)$ we add the transition $(S_{p_1 \ldots p_{n-1}}, p_n, S_{p_2 \ldots p_n})$.

Since $\mathcal{L}(G) \subseteq \mathcal{L}(R)$, then given a query Q, if $[[Q]]_{R,K} = \emptyset$ then it follows that $[[Q]]_{G,K} = \emptyset$. Our optimisation works as follows: For each query Q generated by the rewriting algorithm and each conjunct (x, P, y) of Q we compute a new property path P' that generates strings in $\mathcal{L}(P) \cap \mathcal{L}(R)$. If $\mathcal{L}(P) \cap \mathcal{L}(R)$ is empty, we can discard the query as it will not return any answers; otherwise, we replace the conjunct (x, P, y) with (x, P', y).

To test our graph summarisation optimisation, we have constructed summaries for three datasets, YAGO, LUBM and DBpedia, and have evaluated a range of queries on them. We have observed that the summarisation optimisation is well suited to large sparse datasets that contain a limited number of property labels (such as YAGO or LUBM), and it improves the performance of queries with the APPROX operator as it replaces the _ symbol, which is costly to execute, with a disjunction of specific properties. However, it is not suitable for dense datasets that contain a large number of property labels, such as DBpedia, since the summary generated is too large to be used in practice for optimising queries. Particularly, when computing $\mathcal{L}(P) \cap \mathcal{L}(R)$ the resulting property path query can be too large to be computed.

4   Future Work

As further work we are investigating combining the two optimisation techniques described here, and also a third optimisation technique based on query containment. The query containment optimisation iterates over every query Q generated by the rewriting algorithm at a given cost and checks if there exists any other generated query Q' that contains it, in other words it checks if $[[Q]]_{G,K} \subseteq [[Q']]_{G,K}$ for every G. If that is the case, then Q will not be included in the set of queries to be evaluated. In order to achieve this optimisation a theoretical study of query containment for flexible SPARQL queries is being undertaken. In [1] Chekol et al. have identified the complexity of SPARQL query containment with regular path queries to be $2^{n^2 \log n}$.

Future work might include design of a query optimiser that chooses to apply one or more of these optimisations depending on the query structure and size. For example, single-conjunct queries do not need the pre-computation optimisation as they cannot be split into sub-queries. Similarly, the execution time of queries that contain only one triple pattern may not improve when using the query containment optimisation.

References

1. Chekol, M.W., Euzenat, J., Genevès, P., Layaida, N.: PSPARQL query containment. Research report, Inria Grenoble Rhône-Alpes/LIG Laboratoire d'Informatique de Grenoble (2011)
2. Frosini, R., Calì, A., Poulovassilis, A., Wood, P.T.: Flexible query processing for SPARQL. Semant. Web **8**(4), 533–563 (2017)
3. Poulovassilis, A., Wood, P.T.: Combining approximation and relaxation in semantic web path queries. In: Patel-Schneider, P.F., Pan, Y., Hitzler, P., Mika, P., Zhang, L., Pan, J.Z., Horrocks, I., Glimm, B. (eds.) ISWC 2010. LNCS, vol. 6496, pp. 631–646. Springer, Heidelberg (2010). doi:10.1007/978-3-642-17746-0_40

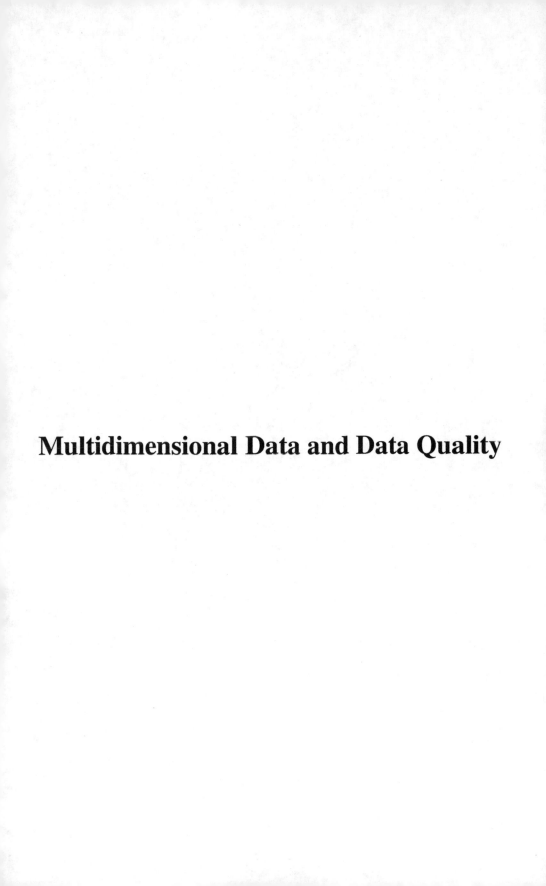

Multidimensional Data and Data Quality

Taming Size and Cardinality of OLAP Data Cubes over Big Data

Alfredo Cuzzocrea[1(✉)], Rim Moussa[2], and Achref Laabidi[3]

[1] DIA Department, University of Trieste and ICAR-CNR, Trieste, Italy
alfredo.cuzzocrea@dia.units.it
[2] LaTICE, ENI-Carthage, University of Carthage, Carthage, Tunisia
rim.moussa@enicarthage.rnu.tn
[3] ENI-Carthage, University of Carthage, Carthage, Tunisia
achref.laabidi@enicarthage.rnu.tn

Abstract. In this paper, we provide three authoritative application scenarios of *TPC-H*d*. The latter is a suitable transformation of *TPC-H* benchmark. The three application scenarios are *(i) OLAP cube calculus on top of columnar relational DBMS*, *(ii) parallel OLAP data cube processing* and *(iii) virtual OLAP data cube design*. We assess the effectiveness and the efficiency of our proposal, using open source systems, namely, *Mondrian* ROLAP server and its OLAP4j driver, *MySQL* - row oriented relational database management system and *MonetDB* -a column-oriented relational database management system.

Keywords: Multidimensional databases · Data warehousing · Schema evolution · Logical OLAP Design

1 Introduction

Decision Support Systems (DSS) are designed to empower the user with the ability to make effective decisions regarding both the current and future activities of an organization. One of the most prominent technologies for knowledge discovery in DSS environments are *On-line Analytical Processing* (OLAP) technologies. OLAP relies heavily upon a data model known as the *multidimensional database* and the *Data cube*. The latter has been playing an essential role in the implementation of *OLAP* [13,29]. However, challenges related to Performance Tuning are to be addressed. Performance Tuning strategies are based on *(i)* adding indexes and summary data for faster data retrieval, *(ii)* data fragmentation such as horizontal partitioning, vertical partitioning, or both in order to enable parallel processing and parallel IO, *(iii)* data clustering for optimizing join-based workloads over relations always queried together, as well as *(iv) hardware scale*, which can be *horizontal scale* for systems allowing data and workload balancing on multiple nodes, and *vertical scale* by operating basic hardware upgrades.

In research experience [9], authors investigated *solutions relying on horizontal data partitioning schemes for parallel building of OLAP data cubes*, and

© Springer International Publishing AG 2017
A. Calì et al. (Eds.): BICOD 2017, LNCS 10365, pp. 113–125, 2017.
DOI: 10.1007/978-3-319-60795-5_12

described the framework *OLAP**, suitable to novel Big Data environments [3, 27], with the specific goal of supporting *OLAP over Big Data* (e.g., [6]), along with the associated benchmark *TPC-H*d*, an appropriate transformation of the well-known data warehouse benchmark *TPC-H* [28] which they used to stress *OLAP**.

With the goal of complimenting previous result [9],in this paper, we overview *TPC-H*d benchmark* in Sect. 2. Then we complement previous research efforts by providing three authoritative application scenarios that build on top of *OLAP**, namely *OLAP Cube Calculus over a relational Columnar-storage* (Sect. 3), *parallel OLAP data cube processing* (Sect. 4) and *virtual OLAP data cube design* (Sect. 5), for which we provide a detailed analysis along with performance evaluation and analysis. We conclude our paper in Sect. 6 by summarizing our research contributions and putting the basis for future work.

2   TPC-H*d Multidimensional Benchmark

The term *On-line Analytical Processing* (OLAP) is introduced in 1993 by E. Codd [5]. This model constitutes a decision support system framework which affords the ability to calculate, consolidate, view, and analyze data according to multiple dimensions. OLAP relies heavily upon *multidimensional databases* (MDB) [2, 4, 12–14, 16, 17, 26, 29]. An *MDB schema* contains a logical model consisting of *OLAP cubes*. Each *OLAP Cube* is described by a *fact table* (facts), a set of *dimensions* and a set of *measures*.

The Transaction Processing Performance Council (TPC) has issued several decision-support benchmarks, including TPC-H benchmark [28]. TPC-H consists of a suite of business oriented adhoc queries and concurrent data modifications. The workload is composed of *(i)* twenty-two parameterized decision-support SQL queries with a high degree of complexity and *(ii)* two refresh functions.

The most important mechanism in OLAP which allow to achieve performance is the use of aggregations. Aggregations are built from the fact table by changing the granularity on specific dimensions and aggregating up data along these dimensions [26]. *TPC-H*d* is a *suitable transformation* of the *TPC-H* benchmark into a *multi-dimensional OLAP benchmark* [8]. Thus, each query of the *TPC-H* workload is mapped onto an OLAP cube, and a temporal dimension (*Time table*) is added to the data warehouse. Also, the *TPC-H* SQL workload translates into a corresponding (*TPC-H*d*) *MDX* workload. The latter implements a client tier which sends a stream of MDX queries in a random order to the database tier, and measures performance of MDX queries for two different workloads. The first workload stream is a *Query Workload*. It is composed of TPC-H queries translated into MDX, while the second is a *Cube-then-Query Workload*. It is composed of TPC-H*d cubes' MDX statements followed by queries' MDX statements. Second workload type should allow query result retrieval from built cubes and consequently, leads to better performance results. *TPC-H*d* design and implementation are detailed [8].

For instance, query Q10 of TPC-H benchmark -*Returned Item Reporting Query* -which SQL statement template is illustrated in Fig. 1, identifies customers who might be having problems with the parts that are shipped to them, and who have returned parts. The query considers only parts that were ordered in a specified quarter of a year. The OLAP cube computes all lost revenues per customer dimension and per order date dimension. The user interacts with the pivot table via an OLAP dice operation, shown in Fig. 2, in order to retrieve lost revenues for French customers during first quarter of 1992.

```
SELECT c_custkey,c_name,c_acctbal, n_name, c_address, c_phone, c_comment,
  SUM(l_extendedprice*(1-l_discount)) as rev
FROM customer, orders, lineitem, nation
WHERE c_custkey = o_custkey  AND l_orderkey = o_orderkey
AND o_orderdate >= date'[DATE]' AND o_orderdate < date'[DATE]' +'3' month
AND l_returnflag = 'R'  AND c_nationkey = n_nationkey
GROUP BY c_custkey,c_name,c_acctbal,c_phone, n_name,c_address,c_comment
ORDER BY revenue desc;
```

Fig. 1. SQL statement of query Q10.

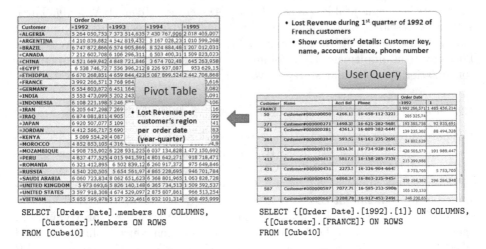

SELECT [Order Date].members ON COLUMNS,
    [Customer].Members ON ROWS
FROM [Cube10]

SELECT {[Order Date].[1992].[1]} ON COLUMNS,
    {[Customer].[FRANCE]} ON ROWS
FROM [Cube10]

Fig. 2. Screenshots of Pivot Tables of C10 and Q10 and corresponding MDX statements.

3 Application Scenario 1: Building OLAP Data Cubes on Top of a Columnar-Oriented DBMS

A column-oriented DBMS is a database management system that stores data tables as columns of data rather than as rows of data [1,22]. The main difference between a columnar database and a traditional row-oriented database are centered around four gains: *(i) high IO performance* with less data moving from hard drives to memory, *(ii) efficient memory management* with only useful data in-memory, *(iii) reduced storage necessities* such that columns with low cardinality and finite domain of definition are compressed, and finally *(iv) efficient schema modifying techniques* such that altering a table schema (p.e. add of a new column) will not induce costly file storage re-organization.

Column-oriented storage layouts are well-suited for OLAP-like workloads. In this section, we present an overview of physical design techniques of column-oriented databases. Then, we report performance measurements of TPC-H*d on top of MonetDB [19] -a columnar relational DBMS.

3.1 Physical Design Techniques for Columnar DBs

Vertical partitioning (VP) of a table T splits T into sub-tables, each of which contains a subset of the columns in T and the primary key or a surrogate key. The key column(s) are required to allow reconstruction of T from its sub-tables. Physical design techniques of column-oriented databases fall into two categories, (i) many design techniques measure the affinity between pairs of attributes through the count of their co-occurrence in the query workload and try to cluster attributes according to their pairwise affinity [15,22] (ii) other techniques represent relational tables using vertical fragmentation, by storing each column in a separate (surrogate key, value) table, called a *Binary Association Table* (BAT) [19]. Three influential research prototypes were developed and pioneered development of commercial systems, which are MonetDB [19], C-Store [25] and VectorWise [30]. These systems implement data columns compression and relational operators optimization for columnar storage layout [1].

3.2 Performance Measurements

The hardware system configuration used for performance measurements are Suno nodes located at Sophia site of GRID5000. Each node has 32 GB of memory, its CPUs are Intel Xeon E5520, 2.27 GHz, with 2 CPUs per node and 4 cores per CPU, and runs Squeeze-x64-xen-1.3 Operating System. Table 1 shows performance measurements for SF = 10 of TPC-H SQL workload, TPC-H*d 'queries MDX Workload and TPC-H*d cube-then-query MDX workload for MySQL -an open-source row-oriented DBMS and MonetDB -an open-source column-oriented DBMS. Performance measurements clearly show that for the same ROLAP server Mondrian, the same data scale and the same workload, MonetDB outperforms MySQL. The reduction in response times for most workload queries is over 90%.

4   Application Scenario 2: Parallel OLAP Data Cube Processing

When dealing with huge data sets, most OLAP systems require high computing capacities and are I/O-bound and CPU-bound. In order to achieve high

Table 1. Performance results for SF = 10, and MySQL and MonetDB as DB backends.

	MySQL DBMS (row-oriented)				MonetDB DBMS (column-oriented)			
	SQL Workload (sec)	MDX Workload (sec)			SQL Workload (sec)	MDX Workload (sec)		
		Query Workload	Cube-then-Query Workload			Query Workload	Cube-then-Query Workload	
			Cube	Query			Cube	Query
Q1	211.94	2,147.33	2,778.49	0.29	2.82	25.6	30.65	0.35
Q2	459.18	1,598.54	346.92	1,565.51	0.2	1,203.5	156.42	1,708.14
Q3	56.75	n/a^{*1}	n/a^{*1}	-	1.46	n/a^{*1}	n/a^{*1}	-
Q4	11.22	1,657.60	7,956.45	5.33	0.69	8.21	8.7	1.2
Q5	19.10	54.53	3,200.64	0.46	1.54	5.07	9,123.98	0.92
Q6	38.63	282.11	371.80	0.53	5.14	5.78	8.38	0.31
Q7	133.92	260.23	617.20	0.06	2.93	3.62	10.7	0.11
Q8	37.18	50.63	2,071.00	4.61	0.72	3.68	35.27	0.5
Q9	645.86	n/a^{*1}	n/a^{*1}	-	1.18	n/a^{*1}	n/a^{*1}	
Q10	191.69	7,100.24	n/a^{*2}	-	1.05	52.5	758.55	17.88
Q11	4.00	2,558.21	3,020.27	1,604.10	0.17	2,536.28	2,834.2	1,313.5
Q12	144.36	456.81	735.67	123.43	2.91	42.78	54.64	23.36
Q13	38.68	n/a^{*2}	n/a^{*2}	-	3.53	n/a^{*2}	n/a^{*2}	-
Q14	122.11	391.06	946.16	0.06	0.19	4.6	18.01	0.09
Q15	90.97	13,005.27	32,064.90	12,413.74	2.9	18.4	532.35	5.4
Q16	47.92	414.82	461.90	4.62	1.36	18.08	64.94	0.99
Q17	4.22	1,131.37	5,711.14	2.03	2.06	8.76	47.48	0.82
Q18	905.16	n/a^{*2}	n/a^{*1}	-	1.68	n/a^{*2}	n/a^{*1}	-
Q19	1.56	598.9	727.72	37.57	0.8	57.22	79.31	0.12
Q20	1.55	14,662.53	n/a^{*3}	-	0.58	423.9	n/a^{*3}	-
Q21	511.54	578.09	855.46	0.15	2.43	5.8	36.74	0.11
Q22	2.40	68.74	402.16	39.33	2.16	100.92	435.8	1.5

- n/a^{*1}: java.lang.OutOfMemoryError: GC overhead limit exceeded,
- n/a^{*2}: java.lang.OutOfMemoryError: Java heap space,
- n/a^{*3}: Mondrian Error:Size of CrossJoin result (200,052,100,026) exceeded limit (2,147,483,647),

performance and large capacity, data management systems rely upon data partitioning, which enables parallel I/Os and parallel processing. In this section, we investigate *solutions relying on data partitioning schemes for parallel building of OLAP data cubes* in a *Shared-Nothing architecture* [12], and we propose the framework *OLAP**, suitable to novel *Big Data environments* [3,27], along with the associated benchmark *TPC-H*d*. We demonstrate through performance measurements the efficiency of the proposed framework, developed on top of the *ROLAP server Mondrian* [23].

4.1 OLAP* Framework Key Considerations for Data Fragmentation

For the design of the data fragmentation schema, we propose the following key considerations.

Reduce the Size of Each Cube to be Built at Each Node: Table 2 lists examples of high-cardinality dimensions in TPC-H*d Benchmark, which imply a high

Table 2. Listing of High-Cardinality Dimensions of TPC-H*d benchmark.

Dimension	Size	OLAP Cubes
Part: p_partkey	SF × 200,000	C2, C11, C21
Customer: c_custkey	SF × 150,000	C10
Supplier: s_suppkey	SF × 10,000	C15, C20
Orders: o_orderkey	SF × 1,500,000	C18
Supplier: s_name	SF × 10,000	C15, C21

cost for buiding an OLAP cube. The data fragmentation scheme should allow parallel building of small cubes through big-cardinality dimensions' partitioning. Consequently, data partitioning will reduce the number of levels to cross in aggregations, and also memory requirements.

Simplify Post-Processing of Parallel OLAP Workload: Every query subject of intra-query parallelism is divided into sub-queries. In order to achieve performance gain, sub-queries run in parallel. Nevertheless, post-processing should be as simple as possible. Three possible strategies of post-processing exist: (*i*) if OLAP cubes have the same dimension members, then the resulting cube has the same dimensions' members as any of the OLAP cube built locally and post-processing consists in adequate processing over locally computed measures; (*ii*) if OLAP cubes have disjoint dimensions' members, then post-processing consists in performing the union of all locally built cubes; finally, (*iii*) if OLAP cubes share dimensions' members, the resulting OLAP cubes is obtained through the merge of all dimensions' hierarchies and performing adequate processing. Notice that the second strategy allows best parallel computing and is the less memory-consuming. In conclusion, the proposed data fragmentation scheme should enable parallel OLAP cube building and implement simple post-processing for federating results obtained from computing nodes.

Enhance Data warehouse Maintenance: Data warehouse maintenance addresses how changes to sources are propagated to a warehouse, including aggregate tables, and OLAP cubes refreshes. The maintenance of the cube should be considered as a bulk incremental update operation. Record-at-a-time updates or full re-computation are not viable solutions.

Reduce Storage Overhead through Controlled Replication: For achieving high performance, and particularly for avoiding inter-sites joins, usually data fragmentation is combined with data replication. Replication allows load balancing (i.e., queries are processed by replicas) and high-availability (i.e., data is available despite some nodes' crash), but has a considerable refresh and storage cost.

4.2   TPC-H*d Fragmentation Schema

Hereafter, we describe *TPC-H*d* horizontally fragmented schema and workload characteristics that OLAP* framework has to deal with,

Table 3. *TPC-H*d* fragmentation schema.

Relation	Schema
Customer	PHPed along *c_custkey*
Orders	DHPed along *o_custkey*
LineItem	DHPed along *Lorderkey*
PartSupp, Supplier, Part, Region, Nation, Time	Replicated

Conflicting Queries Recommendations: In Table 4, we enumerate for each big table of *TPC-H*d* relational schema, the list of partitioning alternatives as well as queries recommending each partitioning schema. Notice that, we distinguish two types of conflicting recommendations, (*i*) conflicting recommendations issued by different queries, such as queries Q4 and Q19; and (*ii*) conflicting recommendations issued by the same query such as for Q16.

High-Cardinality Dimensions: When designing a multi-dimensional database, the number of records in dimensions tables will greatly influence the overall system performance. Some dimensions are huge with millions of members. We call these *High Cardinality Dimensions* (HCD). Table 2 lists HCD within *TPC-H*d* benchmark, and in which OLAP cubes they show up.

Table 3 shows the proposed fragmentation schema of *TPC-H*d* relational data warehouse. We recall that PHP stands for *Primary Horizontal Partitioning* and DHP stands for *Derived Horizontal Partitioning* (e.g., [18,24]), which both are well-known data warehouse partitioning strategies. With respect to the partitioning scheme shown in Table 3, typical queries of TPC-H [28] can be in turn partitioned into three different types of queries. These query classes are the following:

– *Class 1*: as result of replication, queries which involve only replicated tables are executed by any node.
– *Class 2*: queries which involve only co-located partitioned tables are executed on one database node as result of partitioning.

Table 4. *TPC-H*d* workload recommendations.

Table	Alternatives of Fragmentation Schemas
LineItem	• Any fragmentation schema: Q1, Q6, Q15
	• DHP along Orders: Q2, Q3, Q4, Q5, Q7, Q8, Q9, Q10, Q12, Q16, Q18, Q21, Q22
	• DHP along Supplier: Q5, Q7, Q8, Q9, Q20, Q21
	• DHP along Part: Q8, Q9, Q14, Q17, Q19, Q20
Orders	• DHP along Customer: Q8, Q9, Q14, Q17, Q19
PartSupp	• DHP along Supplier: Q2, Q11, Q16
	• DHP along Part: Q2, Q16

- *Class 3*: queries which involve both partitioned and replicated tables are executed on all database nodes. In this class, we distinguish two types of cubes' post-processing, namely:
  - *Sub-Class 3.1*: cubes built at edge servers have completely different dimension members, consequently the result cube is obtained by operating the *UNION ALL* of cubes built at edge servers;
  - *Sub-Class 3.2*: cubes built at edge servers present shared dimension members, consequently the result cube requires operating specific aggregate functions over measures.

4.3   Performance Analysis

Table 5 shows detailed performance results for scale factor equal to SF = 10, comparing *TPC-H*d* workload performances of a single DB back-end to a cluster

Table 5. Performance results of *OLAP* - MySQL/Mondrian* with TPC-H*d benchmark for SF = 10: single DB back-end vs. 4 MySQL DB back-ends.

	MDX Workload (sec)			Parallel MDX Workload (sec)			Parallel+ MDX Workload (sec)		
	Query Workload	Cube-then-Query Workload		Query Workload	Cube-then-Query Workload		Query Workload	Cube-then-Query Workload	
		Cube	Query		Cube	Query		Cube	Query
Q1	2,147.33	2,778.49	0.29	485.73	862.77	0.19	1.10	1.32	0.25
Q2	1,598.54	2,346.92	1,565.51	1,720.2	985.07	1,896.03	n/a^{*1}	n/a^{*1}	-
Q3	n/a^{*1}	n/a^{*1}	-	n/a^{*2}	n/a^{*2}	-	n/a^{*2}	2,106.23	n/a^{*2}
Q4	1,657.60	7,956.45	5.33	523.67	1,657	1.54	0.06	0.07	0.05
Q5	54.53	3,200.64	0.46	12.96	1,219.19	0.19	0.12	0.99	0.06
Q6	282.11	371.80	0.53	72.58	131.70	0.37	0.42	0.77	0.37
Q7	260.23	617.20	0.06	36.01	195.24	0.06	0.08	0.95	0.06
Q8	50.63	2,071.00	4.61	13.38	716.10	2.70	0.07	3.83	0.23
Q9	n/a^{*1}	n/a^{*1}	-	n/a^{*1}	n/a^{*1}	-	n/a^{*2}	n/a^{*2}	-
Q10	7,100.24	n/a^{*2}	-	2,654.20	13,674.02	1,599.47	127.67	9545.68	5.16
Q11	2,558.21	3,020.27	1,604.10	535.75	990.75	505.2	587.99	875.33	497.67
Q12	456.81	735.67	123.43	223.6	467.9	45.7	0.06	0.13	0.06
Q13	n/a^{*2}	n/a^{*2}	-	n/a^{*2}	n/a^{*2}	-	0.08	0.16	0.05
Q14	391.06	946.16	0.06	112.41	356.8	0.05	0.06	0.13	0.05
Q15	13,005.27	32,064.90	12,413.74	2870.56	7,832.22	1,945.7	0.05	0.45	0.03
Q16	414.82	461.90	4.62	640.59	615.77	9.27	3.15	5.25	0.71
Q17	1,131.37	5,711.14	2.03	279.56	1,150.86	2.05	0.10	0.12	0.05
Q18	n/a^{*2}	n/a^{*1}	-	12,331.92	13,111.99	8,272.76	0.02	0.05	0.02
Q19	598.9	727.72	37.57	296.18	330.07	15.57	4.89	6.78	0.45
Q20	14,662.53	n/a^{*3}	-	11,842.90	n/a^{*5}	-	2909.71	n/a^{*6}	-
Q21	578.09	855.46	0.15	185.10	272.39	0.21	2.04	25.12	0.71
Q22	68.74	402.16	39.33	8.19	98.71	13.67	6.7	60.4	3.67

- n/a^{*1}: java.lang.OutOfMemoryError: GC overhead limit exceeded,
- n/a^{*2}: java.lang.OutOfMemoryError: Java heap space
- n/a^{*3}: Mondrian Error:Size of CrossJoin result (200,052,100,026) exceeded limit (2,147,483,647),
- n/a^{*4}: Mondrian Error:Size of CrossJoin result (200,050,000,000) exceeded limit (2,147,483,647),
- n/a^{*5}: Mondrian Error:Size of CrossJoin result (199,940,270,958) exceeded limit (2,147,483,647),
- n/a^{*6}: Mondrian Error:Size of CrossJoin result (199,935,171,300) exceeded limit (2,147,483,647).

composed of 4 MySQL DB back-ends. We also report performance results for $N = 4$, with usage of summary data, as sketched in [20,21], namely,

- *Aggregate tables* for queries Q1, Q3–Q8, Q12–Q20, and Q22. The corresponding OLAP cubes' sizes are scale factor independent or very sparse (e.g. C15 and C18).
- *Derived attributes* for queries Q2, Q9–Q11 and Q21. The corresponding OLAP cubes' sizes are scale factor dependent. For these queries, aggregate tables tend to be very big, and are consequently tuned with derived attributes.

Experiments show that,

- For some queries, cube building is not improving performances such as Q2. The corresponding MDX statement include new members calculus (i.e., measures or named sets), or perform filtering on levels' properties. This constrains the system to build a new pivot table for the query,
- *OLAP** demonstrates good performances for $N = 4$. Indeed, except Q3 and Q9 for which, the system was unable to run MDX statements for memory leaks, the rest of queries were improved through parallel cube building.
- Response times of queries of both workloads, for which aggregate tables were built, namely Q1, Q3–Q8, Q12–Q20, and Q22, were improved. Indeed, most cubes are built in a fraction of a second, especially for those which corresponding aggregate tables are small (refer to Table 6 for aggregate tables' sizes).
- The impact of derived attributes is mitigated. Performance results show good improvements for Q10 and Q21, and small impact on Q11.
- The calculus of aggregate tables is reported in Table 6. Overall, aggregate tables built time is improved, except for those which involve exclusively not fragmented tables.

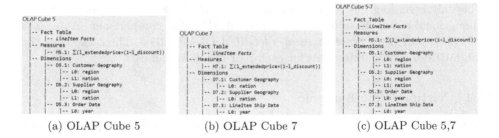

(a) OLAP Cube 5   (b) OLAP Cube 7   (c) OLAP Cube 5,7

Fig. 3. Example of merge of OLAP cubes.

Table 6. Aggregate tables building times (in seconds) for *TPC-H*d* with SF = 10: *single DB back-end* vs. *4 DB back-ends*, for *MySQL* and *MonetDB* DBMS.

	#Rows	MySQL DBMS			MonetDB DBMS		
		Volume	1 DB (s)	4 DBs (s)	Volume	1 DB (s)	4 DBs (s)
agg_C1	129	16.62 KB	343.91	71.63	7.3 KB	3.5	1.01
agg_C3	2,210,908	103.32 MB	173.45	39.52	52.71 MB	4.2	1.5
agg_C4	135	5.22 KB	138.45	32.92	2.24 KB	5.4	1.3
agg_C5	4,375	586.33 KB	822.29	198.6	111 KB	17.5	3.3
agg_C6	1,680	84.67 KB	148.29	42.67	52.5 KB	6.6	1.7
agg_C7	4,375	372.70 KB	720.26	187.8	40.1 KB	29.7	7.9
agg_C8	131,250	12.77 MB	2894.38	818.82	2.13 MB	40.6	9.8
agg_C12	49	3.15 KB	186.68	43.94	1.38 KB	1.01	0.29
agg_C14	84	6.33 KB	367.88	146.76	1.96 KB	4.7	1.2
agg_C15	28	3.84 KB	10,904	852.84	0.71 KB	46.6	16.1
agg_C16	187,495	10.03 MB	63.05	62.26	4.29 MB	3.5	3.5
agg_C17	1,000	45.92 KB	3,180.26	435.52	19.53 KB	991	108
agg_C18	624	37.56 KB	905.16	212.32	18.28 KB	6.8	1.4
agg_C19	854,209	80.65 MB	88.57	26.10	24.44 MB	138.2	25.1
agg_C22	25	1.73 KB	6.25	1.15	0.48 KB	2.7	9.35

5   Application Scenario 3: Virtual OLAP Data Cube Design

Having multiple and small cubes results in faster query performance than one big cube. Nevertheless, it induces additional storage cost and CPU computing if the workload is run against OLAP cubes having same fact table and multiple shared dimensions. A virtual cube represents a subset of a physical cube. *Virtual Cubes* are recommended for minimal maintenance cost of OLAP cubes. They allow finding out shared and relevant materialized pre-computed multidimensional cubes. In order to automate OLAP cubes comparisons, *AutoMDB* [10] parses an XML description of *TPC-H*d* OLAP cubes. Then, sketches the similarities and the differences for each pair of OLAP cubes.

5.1   Virtual Cube Example

AutoMDB [10] allows finding out shared and relevant materialized pre-computed multidimensional cubes. Then, it recommends merge of OLAP cubes based on maximum shared properties and minimum different properties. For instance, *AutoMDB* detects that, (*i*) OLAP cubes *C5* and *C7* have the same fact table which is LINEITEM table. Moreover, (*ii*) both cubes calculate the same measure $sum(l\_extendedprice \times (1 - l\_discount))$, and (iii) two dimensions of OLAP cube *C7* could be collapsed within dimensions of OLAP cube *C5*. Figure 3, illustrates both dimensions sets of OLAP cubes *C5* and C7, as well cube *C_5_7* resulting from the merge of cubes *C5* and *C7*.

Notice that both OLAP cubes *C5* and *C7* sizes are equal to 4375 ($25 \times 25 \times 7$), respectively the *number of customer's nations* multiplied by the *number of suppliers' nations* multiplied by the *number of orders' dates years* for *C5*; and the *number of customer nations* multiplied by the *number of suppliers' nations* multiplied by the *number of line ship date years* for *C7*. Nevertheless, the size of Cube *C_5_7* is equal to 30,625 ($25 \times 25 \times 7 \times 7$), which is the *number of customer's nations* multiplied by the *number of suppliers' nations* multiplied by the *number of orders' dates years* multiplied by the *number of line ship date years*. The size of Cube *C_5_7* is 3.5 times the size of each of *C5* and *C7* OLAP cubes.

5.2 Performance Analysis

Table 7 reports performance measurements conducted for evaluating the cost of merge of OLAP cubes, and running MDX statements related to virtual Cubes *VC5* and *VC7*. Performance measurements show that the time for building *C_5_7* is higher than the ones related to *C5* and *C7*. Nevertheless, building virtual cubes following the physical cube allows a gain in performance whether is the order of execution of cube MDX statements (i.e., C5 then C7 or the inverse).

Table 7. Virtual OLAP cubes (VC5 and VC7) building performances after OLAP cube C_5_7 building.

	Initial schema (sec)	Virtual Cubes (sec)
C5	3,200.64	0.7
C7	617.20	0.2
C_5_7	-	3,457.7

6   Conclusions and Future Work

Following research result [9], in this paper we have provided three authoritative application scenarios that build on top of *TPC-H*d* -a multi-dimensional database benchmark, namely *(i) OLAP cube calculus on top of columnar relational DBMS, (ii) parallel OLAP data cube processing* and *(iii) virtual OLAP data cube design*. We have also provided a comprehensive performance evaluation and analysis.

Future work is mainly devoted to two important aspects: (*i*) integrating novel *data cube compression approaches* (e.g., [7,11]) in order to speed-up efficiency; (*ii*) further stressing the fragmentation phase by integrating *intelligent recommenders* for performance tuning.

References

1. Abadi, D., Boncz, P.A., Harizopoulos, S., Idreos, S., Madden, S.: The design and implementation of modern column-oriented database systems. Found. Trends Databases **5**(3), 197–280 (2013)
2. Agarwal, S., Agrawal, R., Deshpande, P., Gupta, A., Naughton, J.F., Ramakrishnan, R., Sarawagi, S.: On the computation of multidimensional aggregates. In: 22th International Conference on Very Large Data Bases, pp. 506–521 (1996)
3. Agrawal, D., Das, S., El Abbadi, A.: Big data and cloud computing: current state and future opportunities. In: 14th International Conference on Extending Database Technology, EDBT, pp. 530–533. ACM (2011)
4. Agrawal, R., Gupta, A., Sarawagi, S.: Modeling multidimensional databases. In: 13th International Conference on Data Engineering, pp. 232–243 (1997)
5. Codd, E.F., Codd, S.B., Salley, C.T.: Providing OLAP (on-line analytical processing) to user-analysts: an IT mandate. Codd Date **32**, 3–5 (1993)
6. Cuzzocrea, A.: Data warehousing and OLAP over big data: a survey of the state-of-the-art, open problems and future challenges. IJBPIM **7**(4), 372–377 (2015)
7. Cuzzocrea, A., Matrangolo, U.: Analytical synopses for approximate query answering in OLAP environments. In: Galindo, F., Takizawa, M., Traunmüller, R. (eds.) DEXA 2004. LNCS, vol. 3180, pp. 359–370. Springer, Heidelberg (2004). doi:10.1007/978-3-540-30075-5_35
8. Cuzzocrea, A., Moussa, R.: Multidimensional database design via schema transformation: Turning TPC-H into the TPC-H*d multidimensional benchmark. In: 19th International Conference on Management of Data, pp. 56–67 (2013)
9. Cuzzocrea, A., Moussa, R.: A cloud-based framework for supporting effective and efficient OLAP in big data environments. In: 14th IEEE/ACM International Symposium on Cluster, Cloud and Grid Computing, pp. 680–684 (2014)
10. Cuzzocrea, A., Moussa, R., Akaichi, H.: AutoMDB: A framework for automated multidimensional database design via schema transformation. In: 19th International Conference on Management of Data, pp. 93–94 (2013)
11. Cuzzocrea, A., Saccà, D., Serafino, P.: A hierarchy-driven compression technique for advanced OLAP visualization of multidimensional data cubes. In: 8th International Conference on Data Warehousing and Knowledge Discovery, pp. 106–119 (2006)
12. DeWitt, D.J., Madden, S., Stonebraker, M.: How to build a high-performance data warehouse (2005). http://db.lcs.mit.edu/madden/highperf.pdf
13. Gray, J., Chaudhuri, S., Bosworth, A., Layman, A., Reichart, D., Venkatrao, M., Pellow, F., Pirahesh, H.: Data cube: a relational aggregation operator generalizing group-by, cross-tab, and sub-totals. J. Data Min. Knowl. Disc. **1**(1), 29–53 (1997)
14. Gyssens, M., Lakshmanan, L.V.S.: A foundation for multi-dimensional databases. In: 23rd International Conference on Very Large Data Bases, pp. 106–115 (1997)
15. Hoffer, J.A., Severance, D.G.: The use of cluster analysis in physical data base design. In: 1st International Conference on Very Large Data Bases, pp. 69–86 (1975)
16. Inmon, W.H.: Building the Data Warehouse. Wiley, New York (2005)
17. Kimball, R., Ross, M.: The Data Warehouse Toolkit: The Definitive Guide to Dimensional Modeling. Wiley, New York (2013)
18. Lima, A.A., Furtado, C., Valduriez, P., Mattoso, M.: Parallel olap query processing in database clusters with data replication. Distrib. Parallel Databases **25**, 97–123 (2009)

19. MonetDB: The column-store pionneer (2015). https://www.monetdb.org/Home
20. Moussa, R.: Massive data analytics in the cloud: TPC-H experience on hadoop clusters. IJWA **4**(3), 113–133 (2012)
21. Moussa, R.: TPC-H benchmark analytics scenarios and performances on hadoop data clouds. In: 4th International Conference on Networked Digital Technologies, pp. 220–234 (2012)
22. Navathe, S.B., Ra, G.M.: Vertical partitioning for database design: a graphical algorithm. In: International Conference on Management of Data, SIGMOD, pp. 440–450. ACM (1989)
23. Pentaho: Mondrian ROLAP Server (2013). http://mondrian.pentaho.org/
24. Stöhr, T., Rahm, E.: Warlock: A data allocation tool for parallel warehouses. In: 27th International Conference on Very Large Data Bases, pp. 721–722 (2001)
25. Stonebraker, M., Abadi, D.J., Batkin, A., Chen, X., Cherniack, M., Ferreira, M., Lau, E., Lin, A., Madden, S., O'Neil, E.J., O'Neil, P.E., Rasin, A., Tran, N., Zdonik, S.B.: C-store: a column-oriented DBMS. In: 31st International Conference on Very Large Data Bases, pp. 553–564 (2005)
26. Surajit, C., Umeshwar, D.: An overview of data warehousing and OLAP technology. In: SIGMOD Record, vol. 26, pp. 65–74. ACM (1997)
27. Thusoo, A., Sarma, J.S., Jain, N., Shao, Z., Chakka, P., Zhang, N., Anthony, S., Liu, H., Murthy, R.: Hive - a petabyte scale data warehouse using hadoop. In: ICDE, pp. 996–1005 (2010)
28. Transaction Processing Council: TPC-H benchmark (2013). http://www.tpc.org
29. Vassiliadis, P.: Modeling multidimensional databases, cubes and cube operations. In: 10th International Conference on Scientific and Statistical Database Management, SSDBM, pp. 53–62 (1998)
30. Zukowski, M., Boncz, P.A.: Vectorwise: beyond column stores. IEEE Data Eng. Bull. **35**(1), 21–27 (2012)

The Ontological Multidimensional Data Model in Quality Data Specification and Extraction

(Extended Abstract and Progress Report)

Leopoldo Bertossi[1($\boxtimes$)] and Mostafa Milani[2]

[1] School of Computer Science, Carleton University, Ottawa, Canada
bertossi@scs.carleton.ca
[2] Department Computing and Software, McMaster University, Hamilton, Canada
mmilani@mcmaster.ca

In this abstract we briefly present, using a running example: (a) the *Ontological Multidimensional Data Model* (OMD model) [3,13] as an ontological, Datalog$^\pm$-based [6] extension of the Hurtado-Mendelzon (HM) model for multidimensional data [8]; (b) its use for quality data specification and extraction via query answering; and (c) some ongoing research.

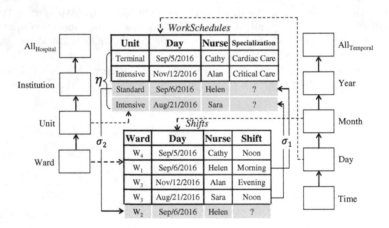

Fig. 1. An OMD model with categorical relations, dimensional rules, and constraints

An OMD model has a *database schema* $\mathcal{R}^\mathcal{M} = \mathcal{H} \cup \mathcal{R}^c$, where $\mathcal{H}$ is a relational schema with multiple dimensions, with a set $\mathcal{K}$ of unary category predicates, and sets $\mathcal{L}$ of binary, child-parent predicates; and $\mathcal{R}^c$ is a set of *categorical predicates*.

Example: Figure 1 shows Hospital and Temporal dimensions. The former's instance is here on the RHS. $\mathcal{K}$ contains predicates $Ward(\cdot)$, $Unit(\cdot)$, $Institution(\cdot)$, etc. Instance $D^\mathcal{H}$ gives them extensions, e.g. $Ward = \{W_1, W_2, W_3, W_4\}$. $\mathcal{L}$ contains, e.g. $WardUnit(\cdot, \cdot)$, with extension: $\{(W_1, \text{standard}), (W_2, \text{standard}), (W_3, \text{intensive}), (W_4, \text{terminal})\}$. In the middle of Fig. 1, *categorical relations* are associated to dimension categories, e.g. $WorkSchedules \in \mathcal{R}^c$. $\square$

© Springer International Publishing AG 2017
A. Calì et al. (Eds.): BICOD 2017, LNCS 10365, pp. 126–130, 2017.
DOI: 10.1007/978-3-319-60795-5_13

Attributes of categorical relations are either *categorical*, whose values are members of dimension categories, or *non-categorical*, taking values from arbitrary domains. Categorical predicates are represented in the form $R(C_1, \ldots, C_m; N_1, \ldots, N_n)$, with categorical attributes before ";" and non-categorical after.

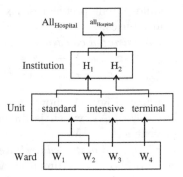

The extensional data, i.e. the instance for the schema $\mathcal{R}^{\mathcal{M}}$, is $I^{\mathcal{M}} = D^{\mathcal{H}} \cup I^c$, where $D^{\mathcal{H}}$ is a complete instance for dimensional subschema $\mathcal{H}$ containing the category and child-parent predicates; and sub-instance I^c contains possibly partial, incomplete extensions for the categorical predicates, i.e. those in $\mathcal{R}^c$. Schema $\mathcal{R}^{\mathcal{M}}$ comes with a set Ω^M of basic, application-independent semantic constraints:

1. Dimensional child-parent predicates must take their values from categories. Accordingly, if child-parent predicate $P \in \mathcal{L}$ is associated to category predicates $K, K' \in \mathcal{K}$, in this order, we introduce inclusion dependencies (IDs) as Datalog$^{\pm}$ *negative constraints (ncs)*: $P(x, x'), \neg K(x) \rightarrow \bot$, and $P(x, x'), \neg K'(x') \rightarrow \bot$. (The $\bot$ symbol is denotes an always false propositional atom.) We do not represent them as Datalog$^{\pm}$'s *tuple-generating dependencies (tgds)* $P(x, x') \rightarrow K(x)$, etc., because we reserve *tgds* for possibly incomplete predicates (in their RHSs).

2. Key constraints on dimensional child-parent predicates $P \in \mathcal{K}$, as *equality-generating dependencies (egds)*: $P(x, x_1), P(x, x_2) \rightarrow x_1 = x_2$.

3. The connections between categorical attributes and the category predicates are specified by means of *ncs*. For categorical predicate R: $R(\bar{x}; \bar{y}), \neg K(x) \rightarrow \bot$, where $x \in \bar{x}$ takes values in category K.

   Example: Categorical predicate *WorkSchedules(Unit,Day;Nurse,Speciality)* has categorical attributes *Unit* and *Day* connected to the Hospital and Temporal dimensions. E.g. the ID *WorkSchedules*[1] $\subseteq$ *Unit*[1] is written in Datalog$^+$ as *WorkSchedules*$(u, d; n, t), \neg Unit(u) \rightarrow \bot$. For the Hospital dimension, one of the two IDs for the child-parent predicate *WardUnit* is *WardUnit*[2] $\subseteq$ *Unit*[1], which is expressed as a *nc*: *WardUnit*$(w, u), \neg Unit(u) \rightarrow \bot$. The key constraint on *WardUnit* is the *egd*: *WardUnit*$(w, u),$ *WardUnit*$(w, u') \rightarrow u = u'$.   □
   The OMD model allows us to build *multidimensional ontologies*, $\mathcal{O}^{\mathcal{M}}$, which contains -in addition to an instance $I^{\mathcal{M}}$ for schema $\mathcal{R}^{\mathcal{M}}$, and the set $\Omega^{\mathcal{M}}$ in 1.-3. above- a set $\Sigma^{\mathcal{M}}$ of *dimensional rules* (those in 4. below), and a set $\kappa^{\mathcal{M}}$ of *dimensional constraints* (in 5. below); of all of them application-dependent and expressed in the Datalog$^{\pm}$ language associated to schema $\mathcal{R}^{\mathcal{M}}$.

4. *Dimensional rules* as Datalog$^+$ *tgds*: $R_1(\bar{x}_1; \bar{y}_1), \ldots, R_n(\bar{x}_n; \bar{y}_n), P_1(x_1, x'_1), \ldots,$ $P_m(x_m, x'_m) \rightarrow \exists \bar{y}' R_k(\bar{x}_k; \bar{y})$. Here, the $R_i(\bar{x}_i; \bar{y}_i))$ are categorical predicates, the P_i are child-parent predicates, $\bar{y}' \subseteq \bar{y}$, $\bar{x}_k \subseteq \bar{x}_1 \cup \ldots \cup \bar{x}_n \cup \{x_1, \ldots, x_m, x'_1, \ldots, x'_m\}$, $\bar{y} \smallsetminus \bar{y}' \subseteq \bar{y}_1 \cup \ldots \cup \bar{y}_n$; repeated variables in bodies (join variables) appear only categorical positions in categorical relations and in

child-parent predicates. Existential variables appear only in non-categorical attributes.

5. *Dimensional constraints*, as *egds* or *ncs*: $R_1(\bar{x}_1; \bar{y}_1), ..., R_n(\bar{x}_n; \bar{y}_n), P_1(x_1, x_1')$, $..., P_m(x_m, x_m') \rightarrow z = z'$, and $R_1(\bar{x}_1; \bar{y}_1), ..., R_n(\bar{x}_n; \bar{y}_n), P_1(x_1, x_1'), ...,$ $P_m(x_m, x_m') \rightarrow \perp$. Here, $R_i \in \mathcal{R}^c$, $P_j \in \mathcal{L}$, and $z, z' \in \bigcup \bar{x}_i \cup \bigcup \bar{y}_j$.

Example: Figure 1 shows a *dimensional constraint* η on categorical relation *WorkSchedules*, which is linked to the Temporal dimension via the *Day* category. It says: *"No personnel was working in the Intensive care unit in January"*: η : *WorkSchedules*(intensive, $d; n, s$), *DayMonth*(d, jan) $\rightarrow \perp$. Figure 1 we also shows the dimensional *tgd* σ_1 : *Shifts*($w, d; n, s$), *WardUnit*(w, u) $\rightarrow$ $\exists t$ *WorkSchedules*($u, d; n, t$), saying that *"If a nurse has shifts in a ward on a specific day, he/she has a working schedule in the unit of that ward on the same day"*. The use of σ_1 generates, from the *Shifts* relation, new tuples for relation *WorkSchedules*, with *null values* for the *Specialization* attribute. Relation *Work Schedules* may be incomplete, and new -possibly virtual- entries can be inserted (the shaded ones showing *Helen* and *Sara* working at the *Standard* and *Intensive* units, resp.). This is done by *upward navigation and data propagation* through the dimension hierarchy.

Now, σ_2: *WorkSchedules*($u, d; n, t$), *WardUnit*(w, u) $\rightarrow \exists s Shifts$($w, d; n, s$) is a dimensional *tgd* that can be used with *WorkSchedules* to generate data for categorical relation *Shifts* (its shaded tuple is one of them). It reflects the guideline stating that *"If a nurse works in a unit on a specific day, he/she has shifts in every ward of that unit on the same day"*. σ_2 supports downward navigation and tuple generation, from the *Unit* category down to the *Ward* category.

If we have a categorical relation *Therm(Ward, Thertype; Nurse)*, with *Ward* and *Thertype* categorical attributes (the latter for an Instrument dimension), the following is an *egd* saying that *"All thermometers in a unit are of the same type"*: *Therm*($w, t; n$), *Therm*($w', t'; n'$), *WardUnit*(w, u), *WardUnit*(w', u) $\rightarrow t = t'$. □

The OMD model goes far beyond classical MD data models. It enables *ontology-based data access* (OBDA) [10] and allows for a seamless integration of a logic-based conceptual model and a relational model, while representing dimensionally structured data. Our MD ontologies have good computational properties [3,13]. Actually, they belong to the class of *weakly-sticky* Datalog$^\pm$ programs [7], for which conjunctive query answering (CQA) can be done in polynomial time in data [14].

With dimensions as fundamental elements of contexts, the OMD model can be use for contextual quality data specification and extraction [13]. The context $\mathcal{O}^c$ is represented as a ontology that contains an OMD model (as a sub-ontology), possibly extra not-necessarily dimensional data, and additional predicate definitions that are used as auxiliary tools for the specification of data quality concerns. A database instance D under quality assessment and quality data extraction is logically mapped into the $\mathcal{O}^c$, for further processing through the context, which provides the otherwise missing elements for addressing or imposing quality concerns on D. The mapping and processing of instance D into/in the context may give rise to alternative quality versions, D^q, of D. (Cf. the figure right below.)

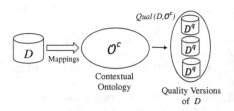

Contextual Ontology

Quality Versions of D

The *quality data* are those shared by all the *quality instances*. Quality data is then extracted from the context through *certain* CQA from the collection, $Qual(D, \mathcal{O}^c)$, of quality instances.

There are several directions of interesting ongoing research in relation to the OMD model and its applications, in general and to data quality. We mention two. First, the interaction of *tgds* and constraints, specially *egds*, may lead to inconsistency. Under certain conditions, such as *separability* [7], the combination with *egds* is computationally easier to handle. In the general case, it may be necessary to apply a *repair semantics*, e.g. to obtain *inconsistency-tolerant* query answers. There are repair semantics for Datalog$^\pm$ ontologies [11] (and also for DL ontologies [9]). Most commonly, the extensional data are (minimally) repaired. In our case, this means repairing MD data, for which certain special MD repair semantics [5,15] may be better than those applied to general relational data [4]. The OMD model allows to express typical MD constraints that guarantee correct *summarizability* (aggregation) [8]. We might also want to keep them satisfied.

Second, the *open-world assumption* on Datalog$^\pm$ (and DL) ontologies makes predicates incomplete (and completable by *tgd* enforcement). Sometimes, in particular with MD extensional data, it may make sense to consider existential variables on categorical attributes with closed domains (e.g. categories and child-parent relations). This creates new issues related to the meaning of existential quantifiers (non-deterministic choices from a fixed set of elements?), data generation, computational aspects, and dimensional navigation (mainly downward, because a parent may have several children). Cf. [2] for Datalog$^\pm$ with closed predicates ([1,12] for the DL case).

References

1. Ahmetaj, S., Ortiz, M., Šimkus, M.: Polynomial datalog rewritings for expressive description logics with closed predicates. In: Proceeding IJCAI 2016, pp. 878–885 (2016)
2. Ahmetaj, S., Ortiz, M., Šimkus, M.: Polynomial datalog rewritings for ontology mediated queries with closed predicates. In: Proceeding of AMW 2016, vol. 1644. CEUR (2016)
3. Bertossi, L. and Milani, M. Ontological multidimensional data models and contextual data quality. Journal submission (2017). Corr ArXiv paper cs.DB/1704.00115
4. Bertossi, L.: Database Repairing and Consistent Query Answering. Synthesis Lectures on Data Management. Morgan & Claypool, San Rafael (2011)
5. Bertossi, L., Bravo, L., Caniupan, M.: Consistent query answering in data warehouses. In: Proceeding of AMW 2009, vol. 450. CEUR (2009)

6. Cali, A., Gottlob, G., Lukasiewicz, T.: Datalog±: a unified approach to ontologies and integrity constraints. In: Proceeding of ICDT 2009, pp. 14–30 (2009)
7. Cali, A., Gottlob, G., Pieris, A.: Towards more expressive ontology languages: the query answering problem. Artif. Intell. **193**, 87–128 (2012)
8. Hurtado, C., Mendelzon, A.: OLAP dimension constraints. In: Proceeding of PODS 2002 (2002)
9. Lembo, D., Lenzerini, M., Rosati, R., Ruzzi, M., Savo, F.: Inconsistency-tolerant query answering in ontology-based data access. J. Web Semant. **3**, 3–29 (2015)
10. Lenzerini, M.: Ontology-based data management. In: Proceeding of AMW 2012, vol. 866. CEUR (2012)
11. Lukasiewicz, T., Martinez, M., Pieris, A., Simari, G.: Inconsistency handling in Datalog+/- ontologies. In: Proceeding of ECAI 2012, pp. 558–563 (2012)
12. Lutz, C., Seylan, I., Wolter, F.: Ontology-based data access with closed predicates is inherently intractable (sometimes). In: Proceeding of IJCAI (2013)
13. Milani, M., Bertossi, L.: Ontology-based multidimensional contexts with applications to quality data specification and extraction. In: Bassiliades, N., Gottlob, G., Sadri, F., Paschke, A., Roman, D. (eds.) RuleML 2015. LNCS, vol. 9202, pp. 277–293. Springer, Cham (2015). doi:10.1007/978-3-319-21542-6_18
14. Milani, M., Bertossi, L.: Extending weakly-sticky Datalog±: query-answering tractability and optimizations. In: Ortiz, M., Schlobach, S. (eds.) RR 2016. LNCS, vol. 9898, pp. 128–143. Springer, Cham (2016). doi:10.1007/978-3-319-45276-0_10
15. Yaghmaie, M., Bertossi, L., Ariyan, S.: Repair-oriented relational schemas for multidimensional databases. In: Proceeding of EDBT (2012)

Distributed and Multimedia Data Management

Closeness Constraints for Separation of Duties in Cloud Databases as an Optimization Problem

Ferdinand Bollwein[1] and Lena Wiese[2(✉)]

[1] Institute of Computer Science, TU Clausthal, Clausthal-Zellerfeld, Germany
`ferdinand.bollwein@tu-clausthal.de`
[2] Institute of Computer Science, University of Goettingen, Goettingen, Germany
`wiese@cs.uni-goettingen.de`

Abstract. Cloud databases offer flexible off-premise data storage and data processing. Security requirements might however impede the use of cloud databases if sensitive or business-critical data are accumulated at a single cloud storage provider. Hence, partitioning the data into less sensitive fragments that are distributed among multiple non-communicating cloud storage providers is a viable method to enforce confidentiality constraints. In this paper, we express this enforcement as an integer linear program. At the same time visibility of certain data combinations can be enabled. Yet in case of violated visibility constraints, the number of different servers on which data is distributed can still be optimized. We introduce novel closeness constraints to express these requirements.

1 Introduction

Cloud databases are a generic tool for outsourcing not only data storage but also data processing: cloud databases offer advanced query and manipulation languages to create database schemas, insert data into tables, query data based on some conditions, update and delete data. Moreover, cloud databases offer joins and aggregation functions. Hence a typical business application of cloud databases is that a cloud customer uploads data into the cloud database and locally only runs scripts to retrieve and manage data on the customer side. This relieves the cloud customer from the burden to install, configure and update a large-scale database system on customer side. Furthermore, depending on changing customer needs, the storage capacity can flexibly be reduced or expanded. However, when private and business-critical data are processed by the cloud database as unencrypted plaintext, cloud database customers have to put a high level of trust in a confidentiality-preserving and privacy-compliant treatment of the data. One way to reduce this trust is to enable the cloud customer to manage distribution of data on as many providers (and under as many user names) as necessary to avoid harmful accumulation of data at a single site.

Separation of duties for cloud databases means that data are split into fragments and these fragments are stored on independent cloud providers. In this paper, vertical fragmentation is used as a technique to protect data confidentiality in cloud databases. Consistently with related work the confidentiality

© Springer International Publishing AG 2017
A. Calì et al. (Eds.): BICOD 2017, LNCS 10365, pp. 133–145, 2017.
DOI: 10.1007/978-3-319-60795-5_14

requirements are modeled as subsets of attributes of the relations. The resulting fragments are explicitly linkable however, it is assumed that they are stored on separate servers which are assumed to be non-communicating. The problem of finding such fragmentations is modeled as a mathematical optimization problem and it is one of the main objectives to minimize the number of servers involved. Moreover, constraints are introduced to improve the usability of the resulting fragmentations and to allow for efficient query answering. Those constraints are modeled as soft constraints in contrast to the confidentiality requirements which are obligatory to be satisfied.

In this paper, we make the following contributions:

- we formalize the enforcement of confidentiality constraints by obtaining *multiple* fragments as a mathematical optimization problem.
- we formalize the distribution of these fragments on *multiple* servers while at the same time minimizing the amount of these servers; that is we obtain a distribution on as few servers as possible.
- moreover, further constraints are introduced to improve the usability of the resulting fragmentations and to allow for efficient query answering; we discuss a weakness of conventional visibility constraints and introduce additional closeness constraints concerned with the distribution of the attributes to allow for efficient query processing.
- visibility and closeness constraints are modeled as soft constraints – in contrast, confidentiality constraints are hard and have to be fully satisfied.

We start this article with a survey of related work in Sect. 2. Section 3 sets the necessary terminology; Sects. 4 and 5 analyze a standard and an extended Separation of Duties problem; Sect. 6 provides a translation into an integer linear program; Sect. 7 briefly describes a prototypical implementation; Sect. 8 concludes the article.

2 Related Work

Horizontal (row-wise) and vertical (column-wise) fragmentation are the two basic approaches to partition tables. Fragmentation as a security mechanism follows the assumption that links between data are highly sensitive (for example, linking a patient name with a disease) whereas individual values (only patient names or only diseases) are less sensitive. The existing approaches can be divided into:

- keep-a-few approaches: some highly sensitive data are maintained at the trusted client side while non-sensitive fragments are stored on an external server (like a cloud database); this approach was pioneered in [5].
- non-communicating servers approaches: fragments are stored on different servers that do not interact; this approach was pioneered in [1].

These approaches only consider fragmentation of a single table into *two* fragments. In the former case (keep-a-few), a server fragment and an owner fragment is obtained; in the latter case (non-communicating servers) two fragments are obtained to be stored on two servers.

In [6] the authors consider multiple fragments however they require these fragments (1) to be unlinkable to be stored on one single external server and (2) to be non-overlapping. In contrast to this, we assume that the servers are non-communicating (and hence allow linkability in particular by tuple ID to enable recombination of results on the client side) and we allow a certain level of overlaps (and hence redundancy of data) to improve data visibility.

Vertical [2] as well as horizontal [14] confidentiality-preserving fragmentations have also been analyzed on a logical background. Last but not least, the article [10] surveys several approaches.

3   Relations and Fragmentation

In this paper we assume the common setting of a database table that has to be vertically split into fragments in order to hide some secret information. A database table consists of a set of columns (the names of which are also called attributes). Each attribute has a data type and an according domain of values denoted $\mathrm{dom}(a_i)$. More formally, we talk about a relation schema $R(A) = R(a_1, \ldots, a_n)$ that consists of a relation name R and a finite set of attributes $A = \{a_1, \ldots, a_n\}$. A *relation* (instance) r on the relation schema $R(a_1, \ldots, a_n)$, also denoted by $r(R)$, is defined as an ordered set of n-tuples $r = (t_1, \ldots, t_m)$ such that each *tuple* t_j is an ordered list $t_j = \langle v_1, \ldots, v_n \rangle$ of values $v_i \in \mathrm{dom}(a_i)$ or $v_i = \mathsf{NULL}$.

In order to enforce confidentiality constraints, we obtain a vertical fragmentation of the table. Each fragment contains a subset of the attributes in A. We have to define a special tuple identifier to be able to recombine the original table from the fragments. More formally, a fragmentation of a table is a set $\mathbf{f} = (f_0, \ldots, f_k)$ of fragments f_i where each f_i contains a tuple identifier *tid* (a candidate key of the relation which can itself consist of several attributes) and further attributes: $f_i = \{tid, a_{i_1}, \ldots, a_{i_k}\}$ where $a_{i_j} \in A$. The fragment f_0 is the dedicated *owner fragment* (which is in particular needed to satisfy the singleton confidentiality constraints); all other fragments $f_1, \ldots, f_m$ are *server fragments* which should be allocated on different non-communicating cloud storage providers. Following [1], due to the non-communication assumption, we allow the different server fragments to be linkable (in particular, by the tuple ID but also be common attributes to achieve higher visibility as described in a later section).

When fragmenting a relation vertically, there are two main requirements. The first one (completeness) is that every attribute must be placed in at least one fragment to prevent data loss. The second property (reconstruction) that must be satisfied is more technical: by including a candidate key in every fragment, it is possible to associate the tuples of the individual table fragments. Equi-join operations on those attributes can then be used to reconstruct the original relation. There is also a third property (disjointness) which is often required. This property demands that every non-tuple-identifier attribute is placed in exactly one vertical fragment. However, especially in the context of this work, there are reasons to omit this property to increase the usability of the resulting vertical fragmentation. Detailed information on this is presented in Sect. 5. Based on this preliminary information, the *correct/lossless vertical fragmentation* of a single relation is formally defined as follows:

Definition 1 (Vertical Fragmentation). *Let r be a relation on the relation schema $R(A)$. Let $tid \subset A$, the tuple identifier, be a predefined candidate key of r. A sequence $\mathbf{f} = (f_0, \ldots, f_k)$ where $f_j \subseteq A$ for all $j \in \{0, \ldots, k\}$ is called a correct vertical fragmentation of r if the following conditions are met:*

- ***Completeness:*** $\bigcup_{j=0}^{k} f_j = A$
- ***Disjointness:*** $f_i \cap f_j \subseteq tid$, *for all* $f_i \neq f_j$ *with* $f_i, f_j \neq \emptyset$
- ***Reconstruction:*** $tid \subset f_j$, *if* $f_j \neq \emptyset$

A fragmentation that satisfies completeness and reconstruction but not necessarily the disjointness property is called a lossless vertical fragmentation *of r.*

The cardinality $\mathrm{card}(\mathbf{f})$ of a correct/lossless vertical fragmentation of r is defined as the number of nonempty fragments of $\mathbf{f}$: $\mathrm{card}(\mathbf{f}) = \sum_{\substack{j=0 \\ f_j \neq \emptyset}}^{k} 1$.

At physical level, the relation fragment *or* table fragment *derived from fragment f_j is given by the projection $\pi_{f_j}(r)$.*

It is further worth noticing that the tuple identifier is required to form a proper subset of the fragments which prohibits fragments consisting of the tuple identifier attributes only. This requirement is due to the fact that the tuple identifier's sole purpose should be to ensure the reconstruction property.

4   Standard Separation of Duties Problem

The security requirements are specified at attribute level, i.e. certain attributes or combinations of attributes are considered sensitive and must not be stored by a single untrusted database server. This can, consistently with related work [1,3–5,11], be modeled with the notion of *confidentiality constraints*.

A *confidentiality constraint* is a subset of attributes of a table: a confidentiality constraint is written as $c \subseteq A$. We differentiate the following two cases:

1. Singleton constraints where the cardinality $card(c) = 1$; that is, c contains only a single attribute $c = \{a_i\}$. In this case, the servers are not allowed to read the values in column a_i.
2. Association constraints (see [10]) with cardinality $card(c) > 1$. In this case, the servers are not allowed to read a combination of values of those attributes contained in the confidentiality constraints. However any real subset of these attributes may be revealed.

Definition 2 (Confidentiality Constraints). *Let $R(A)$ be a relation schema over the set of attributes A. A confidentiality constraint on $R(A)$ is defined by a subset of attributes $c \subseteq A$ with $c \neq \emptyset$. A confidentiality constraint c with $|c| = 1$ is called a singleton constraint; a confidentiality constraint c that satisfies $|c| > 1$ is called an association constraint.*

As an example, consider a table containing information about patients of a hospital. We might have highly sensitive identifying attributes like name and SSN (social security number); these would then be turned into singleton confidentiality constraints. On the other hand, some attributes are only sensitive in combination: the birth year, the ZIP code and the gender in combination can act as a quasi-identifier which can reveal a patient's identity. In this case, any subset of birth year, ZIP code and gender may be revealed but not the entire combination.

As attributes contained in a singleton constraint are not allowed to be accessed by an untrusted server, they cannot be outsourced in plaintext at all. Because we refrain from using encryption those attributes have to be stored locally at the owner side. On the other hand, association constraints can be satisfied by distributing the respective attributes among two or database servers. More precisely, a correct vertical fragmentation $\mathbf{f} = (f_0, \ldots, f_k)$ has to be found in which one fragment stores all the attributes contained in singleton constraints and all other fragments are not a superset of a confidentiality constraint. As a common convention throughout the rest of this work, fragment f_0 will always denote the *owner fragment* which stores all the attributes contained in singleton constraints. This fragment is stored by a local, trusted database. The other fragments $f_1, \ldots, f_k$ denote the *server fragments* and each of those is stored by a different untrusted database server. We require the server fragments $f_1, \ldots, f_k$ to obey a given set of confidentiality constraints $C = \{c_1, \ldots, c_l\}$. A server fragment f_j is confidentiality-preserving if $c \nsubseteq f_j$ for all $c \in C$. This leads to the formal definition of a *confidentiality-preserving vertical fragmentation*:

Definition 3 (Confidentiality-preserving Vertical Fragmentation). *For relation r on schema $R(A)$ and a set of confidentiality constraints C, a correct/lossless vertical fragmentation $\mathbf{f} = (f_0, \ldots, f_k)$ preserves confidentiality with respect to C if for all $c \in C$ and $1 \leq j \leq k$ it holds that $c \nsubseteq f_j$.*

It is necessary to introduce some reasonable restrictions to the set of confidentiality constraints. These restrictions are of theoretical nature and will not restrict its expressiveness. These requirements are summarized by the following definition of a *well-defined* set of confidentiality constraints (we extend the definition of e.g. [10] to our special treatment of tuple identifiers):

Definition 4 (Well-defined Set of Confidentiality Constraints). *Given a relation r on the relation schema $R(A)$ and a designated tuple identifier $tid \subset A$. A set of confidentiality constraints C is well-defined if it satisfies:*

- *For all $c, c' \in C$ with $c \neq c'$, it holds that $c \nsubseteq c'$.*
- *For all $c \in C$, it holds that $c \cap tid = \emptyset$.*

The first condition requires that no confidentiality constraint c is a subset of another confidentiality constraint c'. By the definition of a confidentiality-preserving vertical fragmentation, the satisfaction of c' would be redundant because $c \nsubseteq f_j$ for $j \in \{1, \ldots, k\}$ implies that $c' \nsubseteq f_j$ for $j \in \{1, \ldots, k\}$ if $c \subseteq c'$.

The second condition requires that the tuple identifier attributes are considered insensitive on their own and in combination with other attributes. The tuple identifier's sole purpose is to ensure the reconstruction of the fragmentation by placing it in every nonempty fragment. If, for example, there would be a confidentiality constraint $c \subseteq$ tid, a confidentiality-preserving vertical fragmentation would require that the corresponding tuple identifier attributes cannot be placed in any server fragment. Therefore, every attribute has to be placed in the owner fragment which basically means that the relation cannot be fragmented at all.

Storage space restrictions might also be an important factor for the vertically fragmented relation: the owner and the server fragments may not exceed a databases' capacity. Hence, we assume that there is a weight function that assigns a weight to each subset of attributes $w_r : \mathcal{P}(A) \longrightarrow \mathbb{R}_{\geq 0}$.

It is quite obvious that the cardinality of the confidentiality-preserving fragmentation to be found is a crucial factor for the quality of the fragmentation. Keeping the number of involved server as low as possible will reduce the customer's costs, lower the complexity of maintaining the vertically fragmented relation and also increase the efficiency of executing queries. Therefore, in the following problem statement, the objective is to find a confidentiality-preserving correct vertical fragmentation of minimal cardinality. Additionally, the capacities of the involved storage locations must not be exceeded. Formally, the *(Standard) Separation of Duties Problem* is hence defined as follows:

Definition 5 (Standard Separation of Duties Problem). *For relation r over schema $R(A)$, a well-defined set of confidentiality constraints C, a dedicated tuple identifier tid $\subset A$, a weight function w_r, storage spaces $S_0, \ldots, S_k$ (where S_0 denotes the owner's storage and $S_1, \ldots S_k$ denote the servers' storages) and maximum capacities $W_0, \ldots, W_k \in \mathbb{R}_{\geq 0}$. Find a correct confidentiality-preserving fragmentation $\mathbf{f} = (f_0, \ldots, f_k)$ of minimal cardinality such that the capacities of the storages are not exceeded, i.e. $w_r(f_j) \leq W_j$ for all $0 \leq j \leq k$.*

One should note that in this general formulation the owner fragment can possibly contain all of the attributes if W_0 is sufficiently large. Moreover, in order to solve the problem, one could first assign all attributes in singleton constraints to the owner fragment and afterwards solve the remaining subproblem without singleton constraints. Hence, by considering appropriate values for W_0 one can influence the size of the owner fragment and the overall resulting fragmentation.

5   Extended Separation of Duties Problem

In many scenarios, it is desirable that certain combinations of attributes are stored by a single server or in other words, these combinations are visible on a single server, because they are often queried together. This can be accounted for with the notion of *visibility constraints*:

Definition 6 (Visibility Constraint). *Let $R(A)$ denote a relation schema over the set of attributes A and let r be a relation over $R(A)$. A visibility constraint over $R(A)$ is a subset of attributes $v \subseteq A$. A fragmentation*

$\mathbf{f} = (f_0, \ldots, f_k)$ *satisfies v if there exists $0 \le j \le k$ such that $v \subseteq f_j$. In this case, define* $\mathrm{sat}_v(\mathbf{f}) := 1$ *and* $\mathrm{sat}_v(\mathbf{f}) := 0$ *otherwise. Furthermore, for any set V the number of satisfied visibility constraints is*

$$\mathrm{sat}_V(\mathbf{f}) := \sum_{v \in V} \mathrm{sat}_v(\mathbf{f}).$$

In contrast to confidentiality constraints, the fulfillment of visibility constraints is not mandatory, i.e. confidentiality constraints are *hard constraints* while visibility constraints are *soft constraints*. Roughly speaking, the following extended version of the Separation of Duties Problem aims at finding a confidentiality-preserving vertical fragmentation that minimizes the number of fragments and maximizes the number of satisfied visibility constraints. While there is not much sense in finding a fragmentation that does not satisfy the completeness property, breaking the disjointness property can help to increase the number of satisfied visibility constraints and therefore, in the upcoming problem definition a lossless but not necessarily correct fragmentation will be required.

Although visibility constraints provide a means of keeping certain attributes close together, i.e. on a single server, they are not useful when a certain constraint cannot be satisfied due to some confidentiality constraint. Consider a relation r over the attributes $A = \{\mathsf{PatientID}, \mathsf{DoB}, \mathsf{ZIP}, \mathsf{Diagnosis}, \mathsf{Treatment}\}$ with the dedicated tuple identifier $\mathsf{PatientID}$. Moreover, let a weight function of r be defined by $w_r(a) = 1$ for all $a \in A$. Furthermore, suppose the owner fragment has a capacity of $W_0 = 0$, and there are 3 servers with capacities $W_1 = 2$, $W_2 = 3$ and $W_3 = 2$. For statistical purposes, a visibility constraint $v = \{\mathsf{DoB}, \mathsf{ZIP}, \mathsf{Diagnosis}\}$ is introduced and to preserve the privacy of the patients, the confidentiality constraint $c = \{\mathsf{DoB}, \mathsf{ZIP}\}$ is enforced. However, because $c \subset v$, the visibility constraint cannot be satisfied. Hence, one possible solution to the problem is given by $\mathbf{f} = \{f_0, f_1, f_2, f_3\}$ with: $f_0 = \emptyset$, $f_1 = \{\mathsf{PatientID}, \mathsf{DoB}\}$, $f_2 = \{\mathsf{PatientID}, \mathsf{ZIP}, \mathsf{Treatment}\}$, $f_3 = \{\mathsf{PatientID}, \mathsf{Diagnosis}\}$. Another possible solution is given by the fragmentation $\mathbf{f}' = \{f_0', f_1', f_2', f_3'\}$ with: $f_0' = \emptyset$, $f_1' = \{\mathsf{PatientID}, \mathsf{DoB}\}$, $f_2' = \{\mathsf{PatientID}, \mathsf{ZIP}, \mathsf{Diagnosis}\}$, $f_3' = \{\mathsf{PatientID}, \mathsf{Treatment}\}$. The important thing to notice here is that in $\mathbf{f}$ the attributes in v are spread among three and in $\mathbf{f}'$ among only two servers. As a result, a query for the three attributes DoB, ZIP and $\mathsf{Diagnosis}$ involves three servers for the first fragmentation and only two for the second. Hence, the query will be processed faster for the second fragmentation because on the one hand, the server that stores f_2' can evaluate conditions on both attributes ZIP and $\mathsf{Diagnosis}$ resulting in smaller intermediate results and on the other hand, there is less communication overhead due to the necessity of two servers only. Therefore, it is reasonable to provide constraints to make sure that certain attributes should be distributed among as few servers as possible. Moreover, as in the following problem statement a lossless fragmentation will be required, those constraints can also be used to limit the number of copies of any individual attribute. This introduces an interesting technique to reduce the setup time of a vertical fragmented relation. These so-called *closeness constraints* are defined as follows:

Definition 7 (Closeness Constraint). *Let $R(A)$ denote relation schema over the set of attributes A and let r be a relation over $R(A)$. A closeness constraint over $R(A)$ is a subset of attributes $\gamma \subseteq A$. Let $\mathbf{f} = (f_0, \ldots, f_k)$ be a correct/lossless vertical fragmentation of r, the distribution $\text{dist}_\gamma(\mathbf{f})$ of γ is defined as the number of fragments that contain one of the attributes in γ:*

$$\text{dist}_\gamma(\mathbf{f}) := \sum_{\substack{j=0: \\ f_j \cap \gamma \neq \emptyset}}^{k} 1$$

For any set Γ of closeness constraints, the distribution $\text{dist}_\Gamma(\mathbf{f})$ is defined as the sum of distributions of $\gamma \in \Gamma$.

The following extended problem definition aims at preserving confidentiality by requiring a lossless fragmentation that does not violate any confidentiality constraint. Moreover, the owner's and the servers' capacities must not be exceeded. Furthermore, the minimization of the weighted sum serves three purposes: The summand $\alpha_1 \, \text{card}(\mathbf{f})$ is responsible for minimizing the cardinality of the fragmentation. By subtracting the summand $\alpha_2 \, \text{sat}_V(\mathbf{f})$, each satisfied visibility constraint will lower the overall objective value. Lastly, the distribution of the closeness constraints is minimized by the summand $\alpha_3 \, \text{dist}_\Gamma(\mathbf{f})$. With these explanations, the *Extended Separation of Duties Problem* is defined as follows:

Definition 8 (Extended Separation of Duties Problem). *For relation r over schema $R(A)$, a well-defined set of confidentiality constraints C, a set of visibility constraints V, a set of closeness constraints Γ, a tuple identifier $tid \subset A$, a weight function w_r, storage spaces $S_0, \ldots, S_k$, maximum capacities $W_0, \ldots, W_k \in \mathbb{R}_{\geq 0}$ and weights $\alpha_1, \alpha_2, \alpha_3 \in \mathbb{R}_{\geq 0}$. Find a lossless confidentiality-preserving fragmentation $\mathbf{f} = (f_0, \ldots, f_k)$ of minimal cardinality which satisfies $w_r(f_j) \leq W_j$ for all $0 \leq j \leq k$ such that the following weighted sum is minimized*

$$\alpha_1 \, \text{card}(\mathbf{f}) - \alpha_2 \, \text{sat}_V(\mathbf{f}) + \alpha_3 \, \text{dist}_\Gamma(\mathbf{f}).$$

A reasonable choice for α_1, α_2 and α_3 is presented in the following. The idea is to assign priorities to the three different objectives. In most scenarios, the overall number of necessary servers will have the highest impact on the usability and therefore, minimizing it should have the highest priority. Hence, the desired solution's cardinality should be minimal. The satisfaction of visibility constraints has the second highest priority and therefore, the resulting fragmentation should minimize the cardinality of the fragmentation and the number of satisfied visibility constraints should be maximal among all other confidentiality-preserving fragmentations of minimal cardinality that do not violate the capacity constraints. Finally, among those solutions, the distribution of the closeness constraints should be minimized. This can be achieved by solving the linear inequalities $\alpha_2|V| + \alpha_3(k+1)|\Gamma| < \alpha_1$ and $\alpha_3(k+1)|\Gamma| < \alpha_2$. Solving these inequalities is straightforward and under the assumption that $|V| > 0$ and $|\Gamma| > 0$, one possible solution is given by $\alpha_1 = 1$, $\alpha_2 = \frac{0.9}{2|V|}$ and $\alpha_3 = \frac{0.87}{2(k+1)|V||\Gamma|}$.

Listing 1. Extended Separation of Duties Problem

$$\text{minimize} \quad \alpha_1 \sum_{j=0}^{k} y_j - \alpha_2 \sum_{v \in V} z_v + \alpha_3 \sum_{\gamma \in \Gamma} \sum_{j=0}^{k} \delta_{\gamma j} \tag{1}$$

$$\text{subject to} \quad \sum_{j=0}^{k} x_{ij} \geq 1, \qquad\qquad\qquad a_i \in A \tag{2}$$

$$x_{ij} = y_j, \qquad\qquad\qquad a_i \in \text{tid}, \, j \in \{0, \ldots, k\} \tag{3}$$

$$\sum_{a_i \in A \setminus \text{tid}} x_{ij} \geq x_{i'j}, \qquad\qquad a_{i'} \in \text{tid}, \, j \in \{0, \ldots, k\} \tag{4}$$

$$\sum_{a_i \in A} w_r(a_i) x_{ij} \leq W_j y_j, \qquad j \in \{0, \ldots, k\} \tag{5}$$

$$\sum_{a_i \in c} x_{ij} \leq |c| - 1, \qquad\qquad j \in \{1, \ldots, k\}, \, c \in C \tag{6}$$

$$\sum_{a_i \in v} x_{ij} \geq u_{vj} |v|, \qquad\qquad j \in \{0, \ldots, k\}, \, v \in V \tag{7}$$

$$\sum_{j=0}^{k} u_{vj} \geq z_v, \qquad\qquad\qquad v \in V \tag{8}$$

$$\sum_{a_i \in \gamma} x_{ij} \leq |\gamma| \delta_{\gamma j}, \qquad\qquad \gamma \in \Gamma, \, j \in \{0, \ldots, k\} \tag{9}$$

6   Integer Linear Program Formulation

In this section, the ILP formulation for the Extended Separation of Duties Problems as shown in Listing 1 will be discussed. All variables x_{ij}, y_j, z_v, u_{vj}, $\delta_{\gamma j}$ are binary. In order to identify which fragments should be nonempty, variables $y_0, \ldots, y_k \in \{0, 1\}$ are introduced for the owner fragment f_0 and for each server fragment $f_1, \ldots, f_k$. A value of one indicates that the respective fragment is nonempty. Furthermore, additional binary variables $x_{ij} \in \{0, 1\}$ for each $a_i \in A$ and $j \in \{0, \ldots, k\}$ are used to indicate that attribute a_i is stored in fragment f_j. Additional indicator variables $u_{vj} \in \{0, 1\}$ for all visibility constraints $v \in V$ and all fragments $j \in \{0, \ldots, k\}$ are introduced which are interpreted as follows: If $u_{vj} = 1$, all attributes in v must be stored in fragment f_j. If $u_{vj} = 0$, all attributes in v may be (but do not have to be) stored in this fragment. Moreover indicator variables $z_v \in \{0, 1\}$ are used to indicate that visibility constraint v is satisfied by at least one fragment. This means that z_v can be equal to one if at least one u_{vj} equals one. Moreover, additional variables $\delta_{\gamma j} \in \{0, 1\}$ for all closeness constraints $\gamma \in \Gamma$ and every fragment $j \in \{0, \ldots, k\}$ are necessary to express that fragment f_j contains one or more attributes of γ.

The objective function (1) minimizes the weighted sum stated in Definition 8 in terms of the variables y_j, z_v and $\delta_{\gamma j}$. Because the Extended Separation of Duties Problem only requires a lossless fragmentation, there is no condition that ensures the disjointness property. Constraint (2) ensures the completeness property by requiring that for each $a_i \in A$ there exists at least one j such that x_{ij} equals one. The following Constraint (3) requires that if a fragment is nonempty, it must include the tuple identifier because if $y_j = 1$ all x_{ij} for all $a_i \in$ tid must be equal to one. Conversely, if the fragment should be empty, i.e. $y_j = 0$, no tuple identifier attribute should be placed in the fragment and therefore, x_{ij} must be equal to zero for each $a_i \in$ tid. In the definition of fragmentation, the tuple identifier is required to be a proper subset of each non-empty fragment. This is achieved by Constraint (4) because every tuple identifier attribute $a_{i'} \in$ tid can only be placed in a fragment f_j, i.e. $x_{i'j} = 1$, if there is at least one non-tuple-identifier attribute a_i placed in the same fragment, i.e. $x_{ij} = 1$. Condition (5) has two functions. On the one hand, if fragment f_j should be nonempty and $y_j = 1$, it ensures that the servers capacity W_j is not exceeded. On the other hand, if $y_j = 0$ and f_j should be empty, all x_{ij} for $a_i \in A$ must equal zero and therefore, no attribute can be stored in that fragment. Side constraint (6) makes sure that at most $\text{card}(c) - 1$ attributes contained in a confidentiality constraint are stored in the same server fragment f_j for $j \in \{1, \ldots, k\}$. On the one hand, this ensures that all attributes in a singleton constraint are stored in the owner fragment and on the other hand that no association constraint is violated. Conditions for the visibility constraints are (7) and (8). Each z_v for all $v \in V$ lowers the objective value if $z_v = 1$. Constraint (8) allows $z_v = 1$ only if one of the u_{vj} is is equal to one. However, due to condition (7), a variable u_{ij} can only take a value of one if $x_{ij} = 1$ for all $a_i \in v$. This means that visibility constraint v is satisfied by fragment f_j. Constraint (9) ensures that for each closeness constraint γ and each fragment f_j, the variable $\delta_{\gamma j}$ can only be zero if no attribute $a_i \in \gamma$ is stored in fragment f_j. Therefore, the distribution of γ and the objective value increases for every fragment f_j that contains an attribute in γ.

From an ILP solution, the fragments f_j can be derived by building the sets:

$$f_j := \begin{cases} \{a_i \in A \mid x_{ij} = 1\}, & \text{if } y_j = 1 \\ \emptyset, & \text{else} \end{cases}$$

These fragments then form a correct vertical fragmentation as required in the problem statement.

It should be mentioned further that in some scenarios some visibility or closeness constraints might be more important to satisfy than others. If this is the case, one can simply introduce weights $\beta_v \in (0, 1]$ for all visibility constraints $v \in V$ and weights $\beta_\gamma \in (0, 1]$ for all $\gamma \in \Gamma$ and use the objective function

$$\alpha_1 \sum_{j=0}^{k} y_j - \alpha_2 \sum_{v \in V} \beta_v z_v + \alpha_3 \sum_{\gamma \in \Gamma} \beta_\gamma \sum_{j=0}^{k} q_{\gamma j}$$

in the ILP formulation. This way, visibility constraints with higher weight will contribute more to the minimization of the objective function. Moreover, reducing the distribution of closeness constraints with higher weights is more important than reducing the distribution of closeness constraints with smaller weights.

7   Prototype and Evaluation

We implemented a prototype fragmentation and distribution system (available at http://www.uni-goettingen.de/de/558180.html) based on the IBM ILOG CPLEX solver and PostgreSQL. For testing we set up a TCP-H benchmark (http://www.tpc.org/tpch/) on a single PC equipped with an Intel Xeon E3-1231v3 @3.40 GHz (4 Cores), 32 GB DDR3 RAM and a Seagate ST2000DM001 2 TB HDD with 7200 rpm running Ubuntu 16.04 LTS. The database servers ran in separate, identical virtual machines which are assigned 4 cores and 8 GB of RAM. The virtual machines are running Ubuntu Server 16.04 LTS with an instance of PostgreSQL 9.6.1 installed. We implemented the distributed setting using foreign data wrapper extension postgres_fdw. On the trusted server hosting the owner fragment we created views for the remote server fragments. We ran all 22 queries of the TPC-H benchmark against a non-fragmented local and against the fragmented installation. It turned out that Postgres was not able to process queries Q_{20} and Q_{17} not even in the unfragmented case and we stopped execution after 30 min. Apart from these, for the view-based queries Table 1 shows the execution time (t) in seconds and the slow down (sd) compared to the execution time of the same query on the original database (ot).

Table 1. TPC-H queries (seconds) on fragments (t), unfragmented (ot), slowdown (sd)

Q	1	2	3	4	5	6	7	8	9	10	11	12	13	14	15	16	18	19	21	22
t	41.18	4.699	10.8	18.6	37.0	4.039	11.65	38.58	18.53	12.5	0.58	10.4	11.8	3.765	8.69	2.98	51.0	1.93	79.111	10.58
ot	2.267	0.353	0.861	3.11	0.95	0.291	0.530	1.305	1.652	1.4	0.19	0.457	1.7	0.341	0.66	0.6	5.99	0.65	1708.5	0.534
sd	18.16	13.31	22.99	5.97	38.9	13.88	21.99	29.57	11.22	8.85	2.98	22.75	6.85	11.04	13.1	4.95	8.51	2.98	0.05	19.81

Overall, the increase in execution time compared to the queries on the non-fragmented database does not follow a specific pattern. The slowdown on the distributed views was always less than 30 times – one query even executed faster on the distributed installation. Execution time hence very much depends on the query plan PostgreSQL establishes. To fully understand what causes the increase in the execution times, one would have to study the execution strategy for each of the queries individually; one could then develop strategies to achieve better performances for queries on the vertically fragmented database.

8   Discussion and Conclusion

We studied the problem of finding a confidentiality-preserving vertical fragmentation as a mathematical optimization problem. To achieve a better distribution of attributes among the servers we introduced closeness constraints in addition to conventional visibility constraints.

In future work, we plan to combine the presented approach with partial encryption of a table similar to several approaches surveyed in [10]. Balancing the amount of encrypted and non-encrypted columns leaves room for further mathematical optimization problems. Moreover combining fragmentation with existing frameworks using novel property-preserving encryption schemes (like in [7–9,12,13]) offers even more options to balance leakage and distribution. Because sensitive associations cannot only occur between columns but also between rows of a database, another interesting extension of this work is to additionally explore horizontal fragmentation (as in [14]) which means that database tables are fragmented and distributed row-wise.

References

1. Aggarwal, G., Bawa, M., Ganesan, P., Garcia-Molina, H., Kenthapadi, K., Motwani, R., Srivastava, U., Thomas, D., Xu, Y.: Two can keep a secret: a distributed architecture for secure database services. In: The Second Biennial Conference on Innovative Data Systems Research (CIDR 2005) (2005)
2. Biskup, J., Preuß, M., Wiese, L.: On the inference-proofness of database fragmentation satisfying confidentiality constraints. In: Lai, X., Zhou, J., Li, H. (eds.) ISC 2011. LNCS, vol. 7001, pp. 246–261. Springer, Heidelberg (2011). doi:10.1007/978-3-642-24861-0_17
3. Ciriani, V., De Capitani di Vimercati, S., Foresti, S., Jajodia, S., Paraboschi, S., Samarati, P.: Fragmentation and encryption to enforce privacy in data storage. In: Biskup, J., López, J. (eds.) ESORICS 2007. LNCS, vol. 4734, pp. 171–186. Springer, Heidelberg (2007). doi:10.1007/978-3-540-74835-9_12
4. Ciriani, V., De Capitani di Vimercati, S., Foresti, S., Jajodia, S., Paraboschi, S., Samarati, P.: Selective data outsourcing for enforcing privacy. J. Comput. Secur. **19**(3), 531–566 (2011)
5. Ciriani, V., Capitani di Vimercati, S., Foresti, S., Jajodia, S., Paraboschi, S., Samarati, P.: Keep a few: outsourcing data while maintaining confidentiality. In: Backes, M., Ning, P. (eds.) ESORICS 2009. LNCS, vol. 5789, pp. 440–455. Springer, Heidelberg (2009). doi:10.1007/978-3-642-04444-1_27
6. Ciriani, V., De Capitani di Vimercati, S., Foresti, S., Jajodia, S., Paraboschi, S., Samarati, P.: Combining fragmentation and encryption to protect privacy in data storage. ACM Trans. Inform. Syst. Secur. (TISSEC) **13**(3), 22 (2010)
7. Popa, R.A., Redfield, C., Zeldovich, N., Balakrishnan, H.: CryptDB: protecting confidentiality with encrypted query processing. In: Proceedings of the Twenty-Third ACM Symposium on Operating Systems Principles, pp. 85–100. ACM (2011)
8. Sarfraz, M.I., Nabeel, M., Cao, J., Bertino, E.: DBMask: fine-grained access control on encrypted relational databases. In: Proceedings of the 5th ACM Conference on Data and Application Security and Privacy, pp. 1–11. ACM (2015)
9. Spillner, J., Beck, M., Schill, A., Bohnert, T.M.: Stealth databases: ensuring user-controlled queries in untrusted cloud environments. In: 8th International Conference on Utility and Cloud Computing, pp. 261–270. IEEE (2015)
10. De Capitani di Vimercati, S., Erbacher, R.F., Foresti, S., Jajodia, S., Livraga, G., Samarati, P.: Encryption and fragmentation for data confidentiality in the cloud. In: Aldini, A., Lopez, J., Martinelli, F. (eds.) FOSAD 2012-2013. LNCS, vol. 8604, pp. 212–243. Springer, Cham (2014). doi:10.1007/978-3-319-10082-1_8
11. De Capitani di Vimercati, S., Foresti, S., Jajodia, S., Livraga, G., Paraboschi, S., Samarati, P.: Fragmentation in presence of data dependencies. IEEE Trans. Dependable Secure Comput. **11**(6), 510–523 (2014)

12. Waage, T., Homann, D., Wiese, L.: Practical application of order-preserving encryption in wide column stores. In: SECRYPT, pp. 352–359. SciTePress (2016)
13. Waage, T., Jhajj, R.S., Wiese, L.: Searchable encryption in apache cassandra. In: Garcia-Alfaro, J., Kranakis, E., Bonfante, G. (eds.) FPS 2015. LNCS, vol. 9482, pp. 286–293. Springer, Cham (2016). doi:10.1007/978-3-319-30303-1_19
14. Wiese, L.: Horizontal fragmentation for data outsourcing with formula-based confidentiality constraints. In: Echizen, I., Kunihiro, N., Sasaki, R. (eds.) IWSEC 2010. LNCS, vol. 6434, pp. 101–116. Springer, Heidelberg (2010). doi:10.1007/978-3-642-16825-3_8

WSN-DD: A Wireless Sensor Network Deployment Design Tool

David Santiago Bonilla Bonilla and Ixent Galpin[(✉)]

Dpto. de Ingeniería, Universidad Jorge Tadeo Lozano, Bogotá, Colombia
david7189@gmail.com, ixent@utadeo.edu.co

Abstract. Query processing techniques have been shown to considerably reduce software development efforts for wireless sensor networks. However, deploying a sensor network remains a difficult and error-prone task. This paper presents a comprehensive tool to assist with the design of sensor network deployments, which takes into account a broader range of concerns than previous approaches. Users specify high-level requirements, such as location of nodes, using a simple graphical interface. The tool invokes a sensor network query processor to statically determine the validity of the deployment set out by the user for a given query workload, estimates performance in terms of the query processor's QoS metrics, and generates executable code.

Keywords: Query processing · Wireless sensor networks · Deployment design · Quality-of-Service

1 Introduction

Following almost two decades of research and development, the use of *wireless sensor networks* (WSNs) has become increasingly pervasive. However, recent accounts of real-word WSN deployments report that deploying a WSN remains a non-trivial and onerous task (see, e.g., [12,17]). The problem that we address in this paper is the design of WSN deployments. We propose WSN-DD (for *Wireless Sensor Network-Deployment Design*), a pre-deployment tool to assist with the design of environmental WSNs, available at http://wsn-dd.utadeo.edu.co/. Given metadata such as WSN characteristics, a data collection task, and *Quality-of-Service* expectations, WSN-DD determines statically the validity of particular set of functional and non-functional requirements. We note that our focus is on WSN deployments whose data collection task is carried out in the context of a *sensor network query processor* (SNQP) (e.g., SNEE [7] or TinyDB [11]).

A diverse range of (often unexpected) problems are encountered during the deployment of a WSN which may lead it to fail [2,3]. Returning to a WSN deployment site to diagnose and resolve issues causes delays and increased costs. In the case of certain deployments, it may be risky to attempt to access WSN nodes once they have been deployed (e.g., as is the case with the landslide monitoring deployment described in [17]) Data loss resulting from WSN failure may also have serious consequences for end-users monitoring a phenomenon.

© Springer International Publishing AG 2017
A. Calì et al. (Eds.): BICOD 2017, LNCS 10365, pp. 146–152, 2017.
DOI: 10.1007/978-3-319-60795-5_15

The reasons that WSNs do not work as expected are broad, including issues relating to the WSN nodes themselves (e.g., premature energy depletion or software bugs), issues relating to links between WSN nodes (e.g., network congestion), path issues (e.g., where data source nodes are not connected to a gateway), and *Quality-of-Service* (QoS) issues (e.g., a WSN lifetime falls short of application requirements) [3]. These problems persist in spite of a plethora of complementary approaches proposed to facilitate different aspects of WSN deployment. SNQPs are one such proposal, as they enable WSNs to be programmed using declarative query languages, and aim to alleviate some of the challenges inherent in developing software for WSNs.

Various solutions have been proposed to support WSNs at different stages in the deployment process. Our focus is on the pre-deployment stage, and is complementary to deployment-time tools used to validate newly deployed WSNs in the field (e.g., TASK [4]) and of post-deployment tools used to monitor ongoing WSN deployment health (e.g., Tolle *et al.* [19]). As explained in the survey of related work in Sect. 4, existing pre-deployment techniques tend to focus on piecewise solutions to the WSN deployment problem. Furthermore, these techniques are not integrated with a SNQP, and are therefore unable to reap the benefits that an SNQP provides [7,11].

The main contribution of this paper is a description of a comprehensive tool to assist with the deployment of WSNs that enables its users to specify high-level requirements for a WSN deployment, and to statically verify its feasibility and obtain information about expected performance in terms of QoS metrics (Sect. 2). The use of the tool is illustrated by an example inspired from a real WSN deployment (Sect. 3).

2   System Description

WSN-DD enables users to specify high-level requirements for WSN deployments over a web-based Google Maps graphical interface[1]. The tool is integrated with SNEE [7,8], an SNQP which optimises declarative queries coupled with QoS expectations and metadata about the underlying computing fabric into a *query execution plan* (QEP) and executable code for the TinyOS operating system [10]. WSN-DD therefore covers a broad range of concerns in the WSN design and deployment process, spanning from requirements specification to programming.

WSN-DD comprises seven steps, in which requirements are solicited in a wizard-like fashion, summarised in Fig. 1. The first two steps are preparatory steps: *Area Selection* enables the user to navigate to the appropriate location the map. In the *Obstacle Definition* step, the user draws polygons directly onto the map, to demarcate obstacles that block wireless communications.

In *Stream Creation* the logical extents to be subsequently queried are defined. This enables the generation of a logical schema metadata for SNEE, which specifies the types of sensors that will comprise each sensor stream.

---

[1] The source code is available from our repository at https://github.com/david7189/wsn-dd on an MIT license.

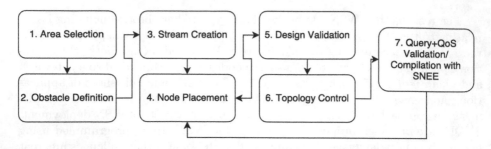

Fig. 1. Summary of the steps in the WSN-DD wizard.

In *Node Placement*, the user adds WSN nodes to the deployment, subject to a pre-defined budget and the cost of each node type. A gateway node, where query results are to be delivered, is also specified. The user associates nodes to the streams defined in the previous step, resulting in a physical schema for SNEE, which maps logical streams to physical WSN nodes.

Design Validation carries out tests to identify connectivity and related issues between WSN nodes. Using the previously defined node locations and obstacles, a network connectivity graph G is derived. Warnings are flagged if (a) source nodes are disconnected from the gateway in G (i.e., there exists no path in G from a source to the gateway), (b) a node is a potential bottleneck (i.e., its degree in G is above a certain threshold), or (c) a node is a single point of failure (i.e., its loss would lead a data source node to become disconnected from G). The intention is that, based on problems flagged, a user will review the node placement in the previous step before proceeding.

In the next step, topology control techniques [18] are employed to simplify the network topology graph G, with the aim of reducing energy consumption and radio interference. The user can select different algorithms to achieve this, viz. Relative Neighbourhood Graph, Delaunay Triangulation and Minimum Spanning Tree [18]. The result of this step is a routing graph G', which a user can visualise before proceeding to the next step. G' constitutes the network description metadata which will be used by SNEE for query validation/compilation.

The final step requests the user to enter a query in SNEEql to specify the functional requirements for a data collection task. The non-functional requirements are expressed in the form of QoS expectations: *acquisition interval* to specify the time between each sensor reading, and *delivery time* which stipulates the maximum time allowed to elapse between a value being acquired at a sensor node, and the respective tuple being delivered at the gateway node. WSN-DD invokes SNEE to determine whether the data collection task embodied by the WSN defined in the previous steps, query and QoS expectations are feasible. If not feasible, the respective error message is given, and the user can make adjustments to the deployment design before retrying. Otherwise, SNEE produces a QEP and executable source code is generated for the WSN nodes. The user is presented with estimated QoS metrics for the QEP such as lifetime, calculated analytically based on cost estimation models [7].

3 Designing an Environmental Monitoring WSN

The *Chingaza Páramo* is a highland ecosystem in the Colombian Andean mountains, critical to the water supply of Bogotá (population: over 8 million). Due to the onset of climate change, it has become the object of close scientific study. The tool was used to aid the design of a WSN deployment carried out by Jorge Tadeo University in Dec 2015 with the aim of monitoring micro-climatic conditions in the Calostros brook area of the ecosystem (although, we note, the deployment itself differed from this example). Each step from Fig. 1 is described in turn. *Step 1*: The user locates the area of interest. *Step 2*: a rocky outcrop identified as a potential obstacle is demarcated as a polygon on the map. *Step 3*: Two streams are defined for deployment, one to monitor flooding of Calostros brook, and another to measure climatic variables, represented by the logical schema:

```
brook(id:int, ts:time, depth:float)
weather(id:int, ts:time, temp:float, humidity:float)
```

Step 4: GPS coordinates taken during a field trip are uploaded and used to specify four points deemed to be of interest for monitoring the brook, where sensing nodes will be located. An additional point on higher ground with a good quality mobile phone signal is assigned as the gateway. The four nodes located along the brook are associated with the **brook** extent, and three nodes (including the gateway) are associated with the **weather** extent. Additional relay nodes are added by clicking on the map to connect the source nodes with the gateway.

Step 5: The topology is validated, and various issues flagged (see Fig. 2). These are rectified by returning to the previous step and adjusting node positions. *Step 6*: The user experiments with different topology control algorithms. *Step 7*: The following query and QoS expectations are entered:

```
SELECT AVG(depth) FROM brook[NOW];
DELIVERY TIME = 3600s, ACQUISITION INTERVAL = 60s
```

The tool reports that the QEP generated is feasible. The user experiments with a shorter delivery time and acquisition interval, which as to be expected, decreases the expected lifetime QoS. As QoS expectations are made more stringent, it is reported that the QEP is no longer feasible.

4 Related Work

The techniques described in the literature to aid WSN deployment design tend to focus on a single step amongst those shown in Fig. 1. Oyman *et al.* [15] focuses on the problem finding an optimal placement of gateway nodes in large-scale

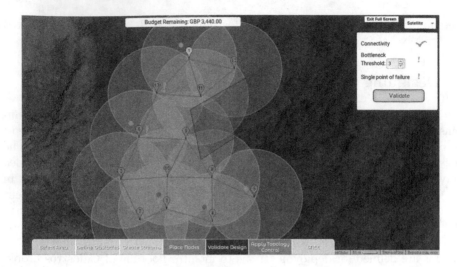

Fig. 2. Screenshot showing WSN deployment under design. The blue polygon is the obstacle defined in *Step 2*; nodes {2, 3, 4, 5} are the sources of the `brook` extent; node 1 is the gateway. The design validation in *Step 5* identifies that nodes {6, 13} are deemed to be single points of failure, and {8, 10} risk being bottlenecks. (Color figure online)

networks, trading off budget and network lifetime. SensDep [16] presents a technique to obtain a node placement trading off coverage and available budget. Mhatre *et al.* [13] present a topology control technique which considers whether single-hop/multi-hop communication is more efficient in terms of energy consumption. We note that in the case of *Node Placement*, optimisation/constraint solving approaches such as [15,16] could be employed to propose an optimal node placement to the user. Similarly, this is the case with [13] for *Topology Control*. However, incorporating such functionality is left as future work.

ANDES [1] aims to find an optimal timing schedule, and to determine whether delivery time QoS can be met, given a node placement and network topology[2]. Tinker [6], Que [5] and Mozumdar *et al.* [14] focus on the design of in-network data processing algorithms for WSNs. Note that, in the case of WSN-DD, this is the responsibility of the SNEE query optimisation step.

WSN-DD was inspired by the tool described in [9] but WSN-DD is more than a reimplementation of [9]. It significantly extends and perfects [9].

5   Conclusions

We have presented WSN-DD, a tool that enables the design of WSN deployments with relative ease. Users of the tool make high-level specifications through

---

[2] In the case of WSN-DD, these concerns are delegated to the SNEE query optimiser, which uses cost models to determine whether meeting a given QoS expectation is feasible [8].

a wizard-like interface and specify a data collection task using the expressive SNEEql query language with associated QoS expectations. WSN-DD statically determines whether a given set of requirements is feasible, and expected WSN performance in terms of QoS metrics, generating executable code for WSN nodes. WSN-DD is the only tool, to our knowledge, that considers such a breadth of issues pertaining to WSN deployment design.

Acknowledgements. We thank Alvaro A.A. Fernandes for comments.

References

1. ANDES: an ANalysis-based DEsign tool for wireless Sensor networks. In: RTSS, pp. 203–213. IEEE (2007)
2. Barrenetxea, G., et al.: The Hitchhiker's guide to successful wireless sensor network deployments. In: SenSys, pp. 43–56. ACM (2008)
3. Beutel, J., et al.: Deployment techniques for sensor networks. In: Ferrari, G. (ed.) Sensor Networks, pp. 219–248. Springer, Heidelberg (2010)
4. Buonadonna, P., et al.: TASK: sensor network in a box. In: EWSN, pp. 133–144. IEEE (2005)
5. Chu, D., Zhao, F., Liu, J., Goraczko, M.: Que: a sensor network rapid prototyping tool with application experiences from a data center deployment. In: Verdone, R. (ed.) EWSN 2008. LNCS, vol. 4913, pp. 337–353. Springer, Heidelberg (2008). doi:10.1007/978-3-540-77690-1_21
6. Elson, J., Parker, A.: Tinker: a tool for designing data-centric sensor networks. In: IPSN, pp. 350–357. ACM (2006)
7. Galpin, I., et al.: SNEE: a query processor for wireless sensor networks. Distrib. Parallel Databases **29**(1), 31–85 (2011)
8. Galpin, I., et al.: QoS-aware optimization of sensor network queries. VLDB J. **22**(4), 495–517 (2013)
9. Hadjipanayis, I.: Designing sensor network deployments. Master's thesis, University of Manchester (2010)
10. Levis, P., et al.: TinyOS: an operating system for sensor networks. In: Weber, W., Rabaey, J.M., Aarts, E. (eds.) Ambient Intelligence, pp. 115–148. Springer, Heidelberg (2005)
11. Madden, S.R., et al.: TinyDB: an acquisitional query processing system for sensor networks. TODS **30**(1), 122–173 (2005)
12. Mafuta, M., et al.: Successful deployment of a wireless sensor network for precision agriculture in Malawi. IJDSN **9**(5), 150703 (2013)
13. Mhatre, V., Rosenberg, C.: Design guidelines for wireless sensor networks: communication, clustering and aggregation. Ad hoc Netw. **2**(1), 45–63 (2004)
14. Mozumdar, M., et al.: A framework for modeling, simulation and automatic code generation of sensor network application. In: SECON, pp. 515–522. IEEE (2008)
15. Oyman, E.I., Ersoy, C.: Multiple sink network design problem in large scale wireless sensor networks. In: ICC, vol. 6, pp. 3663–3667. IEEE (2004)
16. Ramadan, R., et al.: SensDep: a design tool for the deployment of heterogeneous sensing devices. In: DSSNS, 10 pp. IEEE (2006)
17. Rosi, A., et al.: Landslide monitoring with sensor networks: experiences and lessons learnt from a real-world deployment. IJSNet **10**(3), 111–122 (2011)

18. Santi, P.: Topology control in wireless ad hoc and sensor networks. ACM Comput. Surv. (CSUR) **37**(2), 164–194 (2005)
19. Tolle, G., Culler, D.: Design of an application-cooperative management system for wireless sensor networks. In: EWSN, pp. 121–132. IEEE (2005)

A Platform for Edge Computing
Based on Raspberry Pi Clusters

Lorenzo Miori, Julian Sanin, and Sven Helmer$^{(\boxtimes)}$

Faculty of Computer Science, Free University of Bozen-Bolzano, Bolzano, Italy
{lorenzo.miori,julian.sanin,sven.helmer}@unibz.it

Abstract. Small credit-card-sized single-board computers, such as the Raspberry Pi, are becoming ever more popular in areas unrelated to the education of children, for which they were originally intended. So far, these computers have mainly been used in small-scale projects focusing very often on hardware aspects. We want to take single-board computer architectures a step further by showing how to deploy part of an orchestration platform (OpenStack Swift) on a Raspberry Pi cluster to make it a useful platform for more sophisticated data collection and analysis applications located at the edge of a cloud. Our results illustrate that this is indeed possible, but that there are still shortcomings in terms of performance. Nevertheless, with the next generation of small single-board computers that have been introduced recently, we believe that this is a viable approach for certain application domains, such as private clouds or edge computing in harsh environments.

1 Introduction

Single-board computers (SBCs), such as Raspberry Pis (RPis), have received a lot of attention recently. So far, many projects involving these computers have been at a rather small scale, though. For example, students use them for school projects or hobbyists implement hardware for everything from home automation to wildlife monitoring [13]. While there have been attempts at increasing the computational power of SBCs by combining them into clusters (see Sect. 2) and many of the hardware issues faced by early SBC clusters, such as power supply and casing, have been largely solved by now, the work on deploying middleware or orchestration platforms on these clusters is just starting. However, if a cluster does not provide some form of middleware or orchestration, it will be hard to attract potential users, as only few of them will reconfigure, rewrite, or tailor their software to match the specifications of a bare-metal SBC cluster.

Most of the existing projects utilize fairly low-level frameworks, such as the Message Passing Interface (MPI) [7], which basically is a communication protocol. Additionally, the experiments in these projects are conducted from the point of view of high performance computing (HPC). Given the fact that many of the low-cost, power-saving SBCs that are available today show a rather poor performance in terms of raw computing power, focusing on this aspect seems curious to us, as Raspberry Pis will not replace HPC hardware when it comes to

© Springer International Publishing AG 2017
A. Calì et al. (Eds.): BICOD 2017, LNCS 10365, pp. 153–159, 2017.
DOI: 10.1007/978-3-319-60795-5_16

applications involving lots of number crunching. Thus, we want to concentrate on storage applications and (robust) data processing in the context of private cloud environments [2], edge computing [10], and the administration of remote and hard-to-reach (sensor) networks [14], pushing some of the data processing and analysis from central cloud servers to small local computing platforms. The main advantages of these platforms is their low power consumption and the resilience added by the cluster architecture. We can keep the storage server up and running even if some of the individual devices break down and it also allows a much finer granularity when it comes to scaling out the server. This makes it an ideal infrastructure for harsh environments, such as computing platforms for developing countries [8] or laboratories with (ad-hoc) sensor networks in the field [11].

Information on deploying software frameworks and platforms on SBC clusters is few and far between and mostly found in web forums and on discussion boards [3,6]. Inspired by John Dickinson, who successfully deployed OpenStack Swift on a single RPi [6] (more on Swift in Sect. 3), we investigated how to deploy this framework on a cluster of RPis to increase the performance. Scaling these results is not straightforward, an experiment with a setup involving more than twenty USB sticks led to a system not being able to cope with the workload [3]. We demonstrate how to deploy OpenStack Swift on a Raspberry Pi cluster and investigate its performance using the ssbench benchmark (see Sect. 4), showing that there are still some limitations when it comes to deploying middleware or orchestration platforms on an RPi cluster.

2   Related Work

One of the earliest RPi projects is a Beowulf cluster[1] built at the Boise State University consisting of 33 RPis running Arch Linux ARM with a master RPI also having access to a network file system (NFS). The performance was tested by computing the number pi using a Monte Carlo algorithm. Iridis-Pi is a cluster consisting of 64 RPis built at the University of Southampton [5], running Raspbian as the operating system. The LINPACK and HPL (high-performance LINPACK) benchmarks were used to test the raw computing performance of the cluster; for a more practically-oriented benchmark, the distributed file system and map-reduce features of Hadoop were installed. While the system was usable, it was very slow due to lack of RAM [5]. The Pi-Crust cluster, consisting of ten Raspberry Pi 2 model B, went down a similar route, employing the Raspbian operating system and also running the LINPACK benchmark [19]. Cloutier et al. [4] go a step further by not only running LINPACK as a benchmark on their 32 node cluster, but also STREAM, which tests the memory performance. Finally, Schot [15] deploys Hadoop on an eight-node RPi2 cluster. However, all of the work described so far only looks at raw computing performance numbers.

The only work we found that comes close to our approach is the system developed at the University of Glasgow, consisting of 56 RPis [18]. The authors

---

[1] http://coen.boisestate.edu/ece/raspberry-pi/.

state that virtualization architectures, such as Xen, tend to be very memory-hungry and are just in the process of being ported to the ARM processor family. The RPis of the Glasgow cluster are running Raspbian, each node hosting three LXCs (Linux Containers, a light-weight OS-level virtualization). However, no performance numbers are reported.

3 Our RPi Cluster and OpenStack Swift

The Bolzano Cluster. Since its inception in 2013 [1], the RPi cluster in Bolzano has undergone some changes. Here we describe the configuration used for deploying OpenStack Swift: Bobo. One of the biggest changes we made was the casing, as the original one made from metal and plastic turned out to be slightly impractical. The new, portable and modular container (shown in Fig. 1) is made from wood and can house 32 to 40 nodes, a network switch, and the power supply. The RPis are screwed onto several wooden trays that can slide out for assembly and maintenance (see Fig. 1(b)). The cases themselves are stackable and a smaller version ("Bobino") is used by the Dresden University of Technology [16].

| (a) Two wooden cases | (b) Internal structure |

Fig. 1. RPi cluster Bobo

Currently, we are refitting some of the cases with Raspberry Pi 2 models; the results presented here were done with the RPi 1 model B. As operating system we use Raspbian, which is a port of the Linux distribution Debian optimized for the ARMv6 instruction set, e.g. providing better performance for floating point arithmetic [12]. With 3 GByte the original system image is quite heavy, so we stripped out the entire GUI stack along with the X server to make it a completely headless system. In this way we freed up more than 1 GByte of memory. Another tweak relates to the BCM 2835 SoC (system on chip). The RAM is shared between the CPU and GPU and, as we are running a headless system, we configured the split so as to allocate the minimal amount possible to the GPU, which is 16 MByte.

For the network configuration we assigned fixed IP addresses to every RPi in the cluster. This not only made our own administration tasks easier, but also simplified the communication between the service endpoints (proxy, load balancer, and other Swift nodes). For remote (root) access we enabled SSH, creating a set of public and private SSL keys.

OpenStack Swift. We decided to use OpenStack, since it is a popular free and open-source software platform for cloud computing. Swift is the storage component of OpenStack and provides a representational state transfer (RESTful) interface (as well as APIs for different platforms), which means that data objects can be easily located and accessed via a URL. Our goal was to scale out the single device solution, which had already been deployed successfully, by adding more nodes to the setup.

While investigating the deployment of OpenStack Swift on Raspberry Pis, we quickly identified the following issues. Running it on a single device does not make much sense, as we lose all the advantages of a replicated system. In addition, every request has to be processed by a single node, which will be completely overwhelmed once the workload increases, as we have no way of balancing the load. We also found that *DevStack*, a popular set of scripts used for deploying an OpenStack environment, has shortcomings when applied in an RPi setting. Even on a regular PC it can take more than ten minutes to set up a full installation on a single machine, on an RPi it is agonizingly slow. Also, it gave us no control over specific Swift parameters, so we could not fine-tune the installation. For these reasons, we chose John Dickinson's script as a starting point (even though it is also lacking some features important to us), modifying and greatly extending it. Our deployment system consists of a set of Bash and Python scripts that are able to set up the whole cluster.[2] Nevertheless, before running the scripts there are some preliminary steps to prepare the system, such as setting up *Memcached* for storing cached requests and authorization as well as *rsync* for replication purposes.

4   Experimental Results and Discussion

Figure 2 shows a few of the results for running the benchmark tool ssbench on our cluster [17]. The workload consists of a mix of create, read, and delete operations sending JSON files with a size between 1 KByte and 2 MByte ("zero byte upload" measures the overhead of transmissions, such as metadata and HTTP directives, without sending any payload).[3] For comparison: a run-of-the-mill laptop had a throughput of 67 operations per second for "zero byte upload" and 16 operations per second for "small test".

Although the performance of the RPi cluster was rather disappointing, we still see room for improvement. In particular, we have identified the following bottlenecks: a relatively weak CPU, a limited network bandwidth, and some

---

[2] Available at https://github.com/unibz-bobo.

[3] Swift distinguishes four different types of nodes: account, container, object, and proxy nodes.

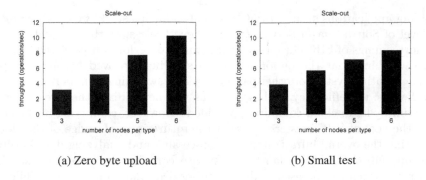

(a) Zero byte upload   (b) Small test

Fig. 2. Scaling out the RPi cluster

issues concerning the file system. We go into more details in the following. The RPi 1 model B we used in our setup employs a 700 MHz ARM processor that achieves roughly 0.04 GFlops (for double precision), which is not a lot compared to CPUs used in laptops or desktops (for instance, an Intel i5-760 quadcore CPU running at 2.8 GHz reaches around 45 GFlops for double precision, while an RPi 2 and 3 (both model B) reach 0.30 GFlops and 0.46 GFlops, respectively, for single precision). One approach for increasing the performance of an RPi is to utilize the GPU, which can provide roughly 24 GFlops. While Broadcom has released the technical (hardware) specification in 2013[4], there is still not a stable API for programming the GPU on RPis.

Currently, RPis use a 100 MBit/s ethernet connection, which is further hampered by being combined with the USB connectors, i.e., all traffic going through the ethernet connection and the USB ports is processed by a combined USB hub/ethernet controller, sharing the USB 2.0 bandwidth of 480 MBit/s. For that reason, the RPis in our cluster used SD cards for storage, as a solid storage disk hooked up to a USB port would have lowered the network bandwidth even further. With some tricks, the ethernet connection can be made faster, for example by using a USB 3.0 Gigabit adapter.[5] However, the gains are limited: with a USB 3.0 Gigabit adapter the speed of the ethernet connection went up to roughly 220 MBit/s. For the Raspberry Pi architecture we are stuck with this bandwidth, an alternative would be to switch to a different platform. For example, the Odroid XU4 offers USB 3.0 and a Gigabit ethernet connector, albeit at a higher energy consumption.

5 Conclusion and Outlook

We see deploying OpenStack Swift on an RPi cluster as a case study, showing how SBC clusters can potentially move from a bare-metal Infrastructure as a Service (IaaS) architecture to a Platform as a Service (PaaS) architecture. By

---

[4] http://www.broadcom.com/docs/support/videocore/VideoCoreIV-AG100-R.pdf.
[5] http://www.jeffgeerling.com/blogs/jeff-geerling/getting-gigabit-networking.

installing applications such as ownCloud on top of this stack, we can even reach the level of Software as a Service (SaaS). Nevertheless, at the moment we still see shortcomings of SBC clusters when it comes to performance-related aspects. While we think that the performance can still be improved by, for example, tuning operating system or other parameters and switching to newer, more powerful devices, the efficiency of these systems will never reach the level of server farms run by large cloud computing providers. However, competing in this area is not the goal of SBC clusters: we see their future in at the edge of the cloud, supporting the overall infrastructure by processing and analyzing data locally.

In our future work we plan to investigate container architectures, such as Docker [9], which can be used to distribute services and applications. While the development of orchestration framework for Docker has started in the form of Kubernetes and Mesos, there are still a lot of open questions when it comes to deploying these frameworks on SBCs.

References

1. Abrahamsson, P., Helmer, S., Phaphoom, N., Nicolodi, L., Preda, N., Miori, L., Angriman, M., Rikkilä, J., Wang, X., Hamily, K., Bugoloni, S.: Affordable and energy-efficient cloud computing clusters: the Bolzano Raspberry Pi cloud cluster experiment. In: UNICO Workshop at CloudCom, Bristol (2013)
2. Basmadjian, R., De Meer, H., Lent, R., Giuliani, G.: Cloud computing and its interest in saving energy: the use case of a private cloud. JoCCASA **1**(1), 1–25 (2012)
3. Bunch, C.: OpenStack Swift, Raspberry Pi, 23 USB keys - aka GhettoSAN v2 (2013). http://openstack.prov12n.com/openstack-swift-raspberry-pi-23-usb-keys-aka-ghettosan-v2/. Accessed Feb 2014
4. Cloutier, M.F., Paradis, C., Weaver, V.M.: Design and analysis of a 32-bit embedded high-performance cluster optimized for energy and performance. In: Co-HPC 2014, pp. 1–8, New Orleans (2014)
5. Cox, S.J., Cox, J.T., Boardman, R.P., Johnston, S.J., Scott, M., O'Brien, N.S.: Iridis-Pi: a low-cost, compact demonstration cluster. Clust. Comput. **17**(2), 349–358 (2014)
6. Dickinson, J.: OpenStack Swift on Raspberry Pi (2013). http://programmerthoughts.com/openstack/swift-on-pi/. Accessed Feb 2014
7. Gropp, W., Lusk, E., Doss, N., Skjellum, A.: A high-performance, portable implementation of the MPI message passing interface standard. Parallel Comput. **22**(6), 789–828 (1996)
8. Helmer, S., Pahl, C., Sanin, J., Miori, L., Brocanelli, S., Cardano, F., Gadler, D., Morandini, D., Piccoli, A., Salam, S., Sharear, A.M., Ventura, A., Abrahamsson, P., Oyetoyan, T.D.: Bringing the cloud to rural and remote areas via cloudlets. In: ACM DEV 2016, Nairobi (2016)
9. Merkel, D.: Docker: lightweight Linux containers for consistent development and deployment. Linux J. **2014**(239) (2014)
10. Pahl, C., Lee, B.: Containers and clusters for edge cloud architectures - a technology review. In: FiCloud 2015, Rome, pp. 379–386, August 2015
11. Porter, J.H., Nagy, E., Kratz, T.K., Hanson, P., Collins, S.L., Arzberger, P.: New eyes on the world: advanced sensors for ecology. BioScience **59**(5), 385–397 (2009)

12. Raspbian.org: Raspbian FAQ. http://www.raspbian.org/RaspbianFAQ. Accessed June 2013
13. Robinson, A., Cook, M.: Raspberry Pi Projects. Wiley, Chichester (2014)
14. Ruponen, S., Zidbeck, J.: Testbed for rural area networking - first steps towards a solution. In: AFRICOMM 2014, Yaounde, pp. 14–23, November 2012
15. Schot, N.: Feasibility of Raspberry Pi 2-based micro data centers in big data applications. In: 23rd Twente Student Conference on IT, Enschede, June 2015
16. Spillner, J., Beck, M., Schil, A., Bohnert, T.M.: Stealth databases: ensuring user-controlled queries in untrusted cloud environments. In: UCC 2015, Limassol, pp. 261–270, December 2015
17. SwiftStack: swiftstack/ssbench. https://github.com/swiftstack/ssbench. Accessed May 2014
18. Tso, P., White, D., Jouet, S., Singer, J., Pezaros, D.: The Glasgow Raspberry Pi cloud: a scale model for cloud computing infrastructures. In: CCRM 2013, Philadelphia (2013)
19. Wilcox, E., Jhunjhunwala, P., Gopavaram, K., Herrera, J.: Pi-crust: a Raspberry Pi cluster implementation. Technical report, Texas A&M University (2015)

Effective Document Fingerprinting

Bettina Fazzinga[1,2], Sergio Flesca[1,2], Filippo Furfaro[1,2], and Elio Masciari[1,2(✉)]

[1] DIMES, University of Calabria, Rende, Italy
{flesca,furfaro}@dimes.unical.it
[2] ICAR-CNR, Rende, Italy
{bettina.fazzinga,elio.masciari}@icar.cnr.it

Abstract. Watermarking digital content is a very common approach leveraged by creators of copyrighted digital data to embed fingerprints into their data. The rationale of such operation is to mark each copy of the data in order to uniquely identify it. These watermarks are embedded in a suitable way to prevent their stripping or modification by users for illegal distribution of the copy. If a copy is illegally distributed by a pirate user (or a set of users referred as coalition) it can be identified by the distributor that can analyze the fingerprint and accuse as traitor the person in charge of that copy. Actions can then be taken against this user, to prevent further illegal distribution. Many approaches have been defined to obtain optimal fingerprinting code based on the well-known Tardos encoding [6]. In this work, we discuss a simple yet powerful decoding scheme that shows effective in many practical case.

1 Introduction

Protection of copyrighted contents is a crucial activity for digital content producers in order to avoid unauthorized use of the artifacts or worse in order to prevent sensible information to be stealth (e.g. private documents of an administrative board). A common solution is the unique identification of each copy by embedding some distinguishing (and not duplicate) features. This activity is usually known as *fingerprinting* and the embedded content is referred as *code*.

In order to make this process robust against possible malicious users attacks, it is mandatory to hide the positions where the code is embedded. Indeed, attacks can be performed by group of malicious users (referred in what follows as pirates), who compare their copies and identify the positions where they differ as a position of the embedded code. The latter is referred as collision attack. If the coalition succeeds in this identification process, pirates can then arbitrarily change the code in these positions. For the purpose of designing proper protection strategies, we can assume however that they do not notice the positions of the hidden code where the bits of their codes agreed and therefore they cannot alter these positions. This assumption is referred as the *marking condition*.

A (collusion resistant) fingerprinting code can be built by a randomized procedure to choose codewords (the code generation) and a tracing algorithm tailored for tracing one of the pirates based on all these codewords and the forged codeword read from the unauthorized copy made by the pirates.

© Springer International Publishing AG 2017
A. Calì et al. (Eds.): BICOD 2017, LNCS 10365, pp. 160–164, 2017.
DOI: 10.1007/978-3-319-60795-5_17

Obviously, we should avoid two type of errors: (1) accusing an innocent user and (2) not accusing a pirate. In this respect, the tracing algorithm fails if it falsely accuses an innocent user or outputs no accused user at all. The above mentioned errors should occur with small probability.

This problem have been largely investigated in the literature as will be discussed in the related work section, and all the approaches proposed so far shares a common terminology that we introduce here in order to ease the reading of next sections.

More in detail, we briefly recall the following key terms:

- *Alphabet size.* The codewords are sequences over a fixed alphabet Σ. Usually, fingerprinting codes are built by leveraging the binary alphabet $\Sigma = \{0, 1\}$, however larger alphabets can be used thus the size Σ of the alphabet is an important parameter;
- *Codelength.* This parameter refers to the length of the codewords, usually denoted by n;
- *Number of users.* Usually denoted by N, it coincides with the number of codewords.
- *Pirate Coalition Size.* This parameter takes into account the actual size of the coalition that could be lower than the expected one (say it c), in such a case, the accusation algorithm should achieve a small error probability[1];
- *Error probability.* A code is ϵ-secure against a coalition of c pirates if the probability of the error of the accusation algorithm is at most ϵ for any set of at most c pirates performing an arbitrary pirate strategy to produce the forged codeword under the marking assumption;
- *Code rate.* The rate R of a fingerprinting code is computed as $R = \frac{log(N)}{n}$, where the logarithm is binary.

The goal of fingerprinting schemes is to find efficient and secure fingerprinting codes while taking into account the high cost of embedding every single digit of the code. This implies that fingerprinting codes should be short.

Tardos fingerprinting [6] is optimum as the code length that is sufficient to guarantee an innocent safety bounded by ϵ that is asymptotically minimum. Moreover, the accusation algorithm allows to detect traitors by looking only at the code they have been assigned to disregarding both the codes assigned to other users and the type of attack that have been performed. It is worth noticing that in literature have been defined many other approaches that outperforms Tardos scheme while having the same asymptotical complexity [4,5]. Unfortunately, Tardos based fingerprinting are not effective in accusation processes when the leveraged code is too short.

In order to overcome the above mentioned limitations, new approaches have been proposed and one of the most interesting is *joint-decoding* Joint-decoders compute the guilty probability for a *set* of users instead of a single one. A first

---

[1] Many fingerprinting algorithm guarantee a small probability of accusing an innocent even if the number of pirates is greater than the expected one. However, in that case, the probability of producing no accusation increases.

proposal has been made in [2,3], however those algorithms are tailored for small coalition and do not scale-up properly. This drawback occurs as the search space computation grows up exponentially w.r.t. the number of users (or the maximum expected number of users that we may conjecture that could form a coalition for spreading the pirated copy).

To ameliorate this problem, joint-decoding have been investigated from the theoretical viewpoint in order to define efficient approaches that work properly for real life situations. In this respect, a *Markov Chain Monte Carlo (MCMC)* based approach has been proposed in [1].

2   A MH-Sampler Based Joint Decoder

Metropolis-Hastings (MH) algorithm aims at approximating a probability density function $F(x_1, \ldots, x_n)$, named target distribution, whose exact formulation is unknown, exploiting the knowledge of a computable function $P(x_1, \ldots, x_n)$, named proposal distribution, that is proportional to $F(x_1, \ldots, x_n)$. The result of an executions of a MH sampler is a sequence of samples. This sequence of samples, which is typically represented as an histogram, yields an approximation of $F(x_1, \ldots, x_n)$ as it is generated with the guarantee that the occurrences of each sample $s \in S$ are proportional to $P(x_1, \ldots, x_n)$ (thus to $F(x_1, \ldots, x_n)$). In a sense, S has a similar shape to F.

Inspired by MH algorithm, we design an accusation algorithm that is reported below.

Algorithm 1. The MH accusation algorithm

Input: A pirated copy y, the code matrix X, the probability array p, an integer $accU$, an integer $cMaxS$
Output: a sequence of k accused users $P = [p_1, \ldots, p_k]$
 1: $cCoal = \emptyset, cPcoal = 0$
 2: $ST = \emptyset$
 3: **for** $i = 1$ to $burnIt + It$ **do**
 4:     $tCoal = cCoal$
 5:     **if** $0 < |tCoal| < cMaxS$ **then**
 6:         $tCoal = genericUpdate(tCoal)$
 7:     **else**
 8:         **if** $|tCoal| = 0$ **then**
 9:             $tCoal = addUser(tCoal)$
10:         **else**
11:             $tCoal = remUser(tCoal)$
12:     $tPCoal = computeProb(tCoal, y)$
13:     $jitter = random()$
14:     **if** $jitter < min(1, \frac{tPCoal}{cPcoal})$ **then**
15:         $cCoal = tCoal, cPcoal = tPCoal$
16:     **if** $i > burnIt$ **then**
17:         $update(ST, cCoal)$
18: **return** $bestPirates(ST, accU)$

Herein: $cCoal$ is the current coalition, $cPcoal$ is the probability of the current coalition, $tCoal$ is the generated coalition and $tPCoal$ is the probability of the generated coalition. Finally, ST is the set of generated coalition (that could contain duplicated elements).

The algorithm works in two phases. The first phase initializes the coalition set by generating $burnIt$ coalitions that will be discarded for the initial tuning of the algorithm. The second phase generates it coalition that are added to ST. Each new coalition is generated by randomly adding or removing a user to the current coalition.

More in detail, function $genericUpdate$ updates the coalition by applying one of the above mentioned operations. Function $addUser$ implements the third operation while $remUser$ implements the second modification strategy. Function $computeProb(tCoal, y)$ computes the probability that a pirated copy y has been generated by coalition $tCoal$ whose users hold the codes included in matrix X_{tCoal} as follows:

$$
computeProb(tCoal, y) = p_{Coal}{}^{|tCoal|-1} \prod_{i=1}^{m} p(X_{tCoal}[i], y[i])
$$

where:

$$
p(X_{tCoal}[i], y[i]) = \begin{cases} \frac{1}{3}, & if\ X_{tCoal}[i]\ contains\ different\ values,\ else \\ p_u, & if\ y[i] = -1 \\ (1 - p_u)p[i], & if\ y[i] = 1\ e\ X_{tCoal}[i] = 1 \\ p_u p[i], & if\ y[i] = 1\ e\ X_{tCoal}[i] = 0 \\ p_u(1 - p[i]), & if\ y[i] = 0\ e\ X_{tCoal}[i] = 1 \\ (1 - p_u)(1 - p[i]), & if\ y[i] = 0\ e\ X_{tCoal}[i] = 0 \end{cases}
$$

The latter implies that the probability that $tCoal$ generates y is assumed to decrease w.r.t. the coalition length by hypothesizing p_{Coal} probability that two users cooperate and we also assume that it is proportional to the probability that the bits of the pirated copy coincides with the coalition ones (under the assumption that they are independent each other).

References

1. Furon, T., Desoubeaux, M.: Tardos codes for real. In: 2014 IEEE International Workshop on Information Forensics and Security, WIFS 2014, Atlanta, pp. 24–29, 3–5 December 2014. http://dx.doi.org/10.1109/WIFS.2014.7084298
2. Nuida, K.: Short collusion-secure fingerprint codes against three pirates. In: Böhme, R., Fong, P.W.L., Safavi-Naini, R. (eds.) IH 2010. LNCS, vol. 6387, pp. 86–102. Springer, Heidelberg (2010). doi:10.1007/978-3-642-16435-4_8
3. Nuida, K., Fujitsu, S., Hagiwara, M., Kitagawa, T., Watanabe, H., Ogawa, K., Imai, H.: An improvement of discrete Tardos fingerprinting codes. Des. Codes Cryptogr. **52**(3), 339–362 (2009). http://dx.doi.org/10.1007/s10623-009-9285-z
4. Skoric, B., Katzenbeisser, S., Celik, M.U.: Symmetric Tardos fingerprinting codes for arbitrary alphabet sizes. Des. Codes Cryptogr. **46**(2), 137–166 (2008). http://dx.doi.org/10.1007/s10623-007-9142-x

5. Skoric, B., Vladimirova, T.U., Celik, M.U., Talstra, J.: Tardos fingerprinting is better than we thought. IEEE Trans. Inf. Theor. **54**(8), 3663–3676 (2008). http://dx.doi.org/10.1109/TIT.2008.926307
6. Tardos, G.: Optimal probabilistic fingerprint codes. J. ACM **55**(2) (2008). www.scopus.com

Author Index

Printed in the United States
By Bookmasters